SIMPLIFIED ROOF FRAMING

SIMPLIFIED ROOF FRAMING

by J. DOUGLAS WILSON

Supervisor, Trade and Industrial Education
Vocational and Practical Arts Branch
Curriculum Division, Los Angeles City Schools

and S. O. WERNER

Associate Professor of Industrial Education and Instructor
of Architectural Drawing University of California,
Santa Barbara College, Santa Barbara, California

SECOND EDITION

Webster Division, McGraw-Hill Book Company

ST. LOUIS NEW YORK SAN FRANCISCO DALLAS
TORONTO LONDON SYDNEY

SIMPLIFIED ROOF FRAMING

70959

20 21 22 23 24 25 26 27 MM 85 84 83 82 81 80 79 78

PREFACE TO THE SECOND EDITION

Continued use of the original text in evening school with skilled carpenters anxious to improve their knowledge through study of more complicated roof work has given opportunity for the development of short-cut methods that have proved of great advantage to the students. This new edition has therefore included a chapter on the steel square, its use as applied to various construction problems, and a description of rafter tables that appear on different makes of squares. Once a carpenter understands the *principles* of roof layout it is easy for him to use the tables.

Chapter VII is new and has been added to aid instructors in giving a review of the roof terms and basic principles of roof framing. A few new drawings have been made to simplify or clarify some of the roof layout procedures. The text has also been carefully reedited and a few changes made.

The continued use of the terms *mathematical* length, *line* length, and *theoretical* length requires a brief explanation. Actual teaching practice and experience have shown that the term line length is most easily understood because it implies no thickness or width of stock but simply a linear measurement. In the text, line length and mathematical length are sometimes used interchangeably.

Acknowledgment is made to Henry Bertch, building contractor, South Pasadena, Calif., and to Ralph Lee, carpentry instructor, Frank Wiggins Trade School, Los Angeles, Calif., for invaluable aid in checking the manuscript and for drawings in the new chapters. Acknowledgment is also made to Stanley Tools, New Britain, Conn.; to Nicholls Manufacturing Co., Ottumwa, Iowa; and to Sargent & Co., New Haven, Conn., for permission to use illustrations of steel framing squares manufactured by these companies.

J. Douglas Wilson
S. O. Werner

PREFACE TO THE FIRST EDITION

Roof framing is a topic that has had considerable attention during the past few years. The number of books already written on the subject attest to its value and one might believe that sufficient had been said concerning it.

The objective of this book, however, is different from that of most of the books now published. While a knowledge of mechanical drawing and geometry is an invaluable aid in the study of roof framing, and these subjects form the foundation upon which much written material is based, yet the subject can be studied and mastered without this background. It is with the thought in mind that many of the skilled mechanics in the building work today do not have a working knowledge of such subjects as mechanical drawing, geometry, or trigonometry that this material has been developed.

The subject matter has been given a thorough try-out in evening school work and has proved successful. It is also applicable to the new field of industrial vocational education which is being so rapidly developed in the modern high school.

For several reasons no mention is made concerning the use of the rafter tables that are stamped on the steel square. First, all steel squares do not have tables stamped on them; second, the tables on the different makes of squares are not the same; third, and of paramount importance, it is the aim of this book to teach *principles* involved in roof framing. If these principles are known, the use of the rafter tables can be quickly learned; if the principles are not known, the worker cannot lay out the rafter intelligently, even though he may be able to "read" the tables.

The material is organized in such a way that each chapter represents a typical roof while each section of each chapter represents a rafter. It is therefore possible to study either a complete roof or simply a rafter, as desired.

As a suggestion to instructors who teach the subject of roof framing, we have found the use of colored chalk, when presenting the drawings on the blackboard, an invaluable aid, as it is possible to call attention in a very definite way to the several lines involved in some of the drawings.

The authors wish especially to acknowledge their indebtedness to the three persons who gave so freely of their time and talent in proof reading the manuscript and drawings both from the mechanical and technical point of view. We refer to Mr. George W. MacKenzie, Head of the Mechanic Arts Department of the Lincoln High School, Los Angeles; Mr. George W. Galbraith, formerly Supervisor of Teacher Training, University of California, Southern Branch, Los Angeles; and Mr. Elmer C. Richardson, Coordinator of the Frank Wiggins Trade School of the same city to whom the writers are indebted for the first group of drawings that helped form the nucleus of the book.

It is the authors' sincere hope that this book will be a means of promoting industrial vocational education in our trade and high schools and that it may be of interest and profit to the mechanic already engaged in the fascinating work of building construction.

J. Douglas Wilson
S. O. Werner

CONTENTS

CHAPTER I

TERMS, TYPES, AND PRINCIPLES

I. Roof Types:

Gable
Hip
Intersecting

II. The Four Layout Problems:

1. Line length
2. Angle of the cut
3. Shape of the cut
4. Shortening

III. Roof Terms:

Span
Unit of span
Total rise
Rise in inches
Unit of run
Total run
Pitch
Cut of the roof
Line length
Plumb and level lines

SIMPLIFIED ROOF FRAMING

CHAPTER I

TERMS, TYPES, AND PRINCIPLES

There are three common types of roofs used in house construction: namely, the gable, hip, and intersecting roofs as illustrated in Fig. 1. Irrespective of the type, the same fundamental principles underlie the framing of all three.

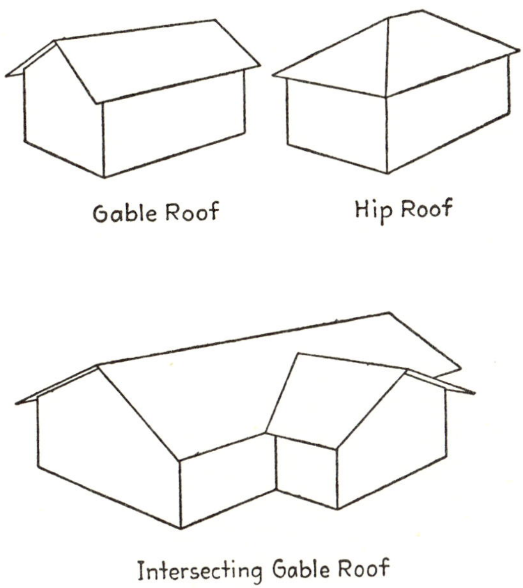

Fig. 1.

The basis of all roof framing is the geometrical figure called a *right-angle triangle*. Figure 2 illustrates such a triangle as applied to a gable roof, and also indicates the roof-framing terms that correspond to the terms of the triangle. The *base* of the triangle becomes the *total run* of the roof; the *altitude* of the triangle becomes the *total rise* of the roof; and the *hypotenuse* of the triangle becomes the *line length* of the common rafter.

All of the problems in roof framing involve mathematical solutions, even though the steel square is used in the layout work. The use of the steel square, however, reduces the mathematical process to a minimum, so that any person of ordinary intelligence, by studying these principles, can learn to frame a roof.

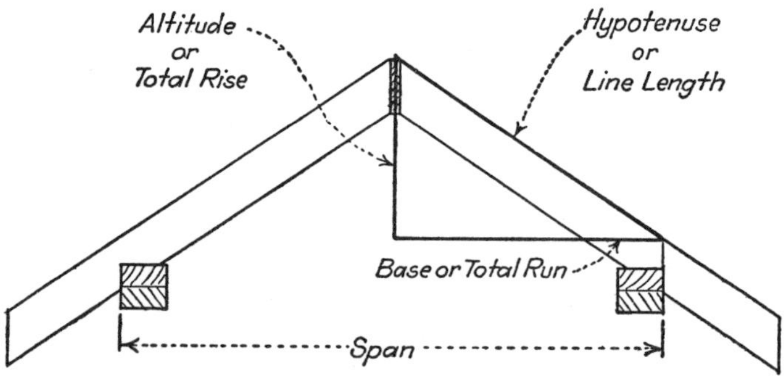

FIG. 2.

There are four problems involved in the layout process of any rafter. They are briefly:

1. Finding the line length of the rafter.
2. Finding the angle of the cuts.
3. Finding the shape of the cuts.
4. Finding the difference between the mathematical and actual length of the rafter. This difference is called "shortening." The principles underlying each of these problems will be developed in the following pages.

ROOF TERMS

A thorough acquaintance with the following terms and definitions is necessary if the science of roof framing is to be mastered.

Span.—(Figure 3.) The span of any roof is the shortest distance between two opposite common rafter seats. Stated in another way, it is the measurement that is at right angles to the direction of the ridge of the building. The span distance is always given on the blue print. Therefore, to find the span of a roof "read the blue print."

Total Rise.—(Figure 3.) The total rise is the vertical distance that the ridge is above the plate line. There are two ways of finding this distance:

1. Measure on the elevation sheet of the blue print.
2. Multiply the span by the pitch of the roof.

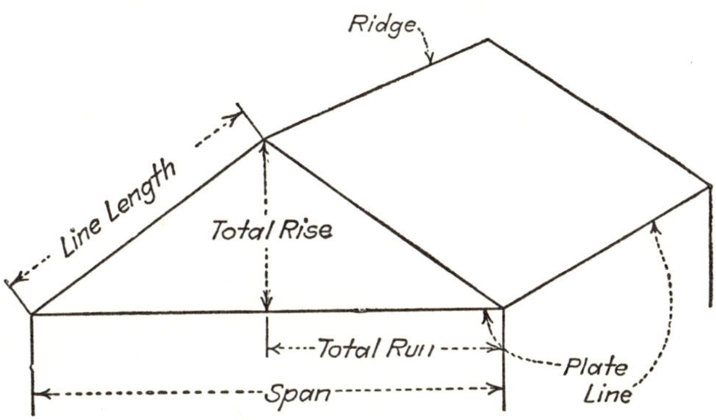

FIG. 3.

To illustrate method 2:

When the span of a roof is 32 ft. and the pitch is ¼, what is the total rise?

Solution: 32 ft. × ¼ equals 8 ft., or, stated as a formula, Span × Pitch = Total rise.

Total Run.—(Figure 3.) The term "total run" always refers to the *level* distance over which any rafter "passes." For the common rafter this would be one-half the span distance.

To illustrate:

What is the total run of a common rafter for a building that has a span of 22 ft.?

Solution: 22 ft. × ½ equals 11 ft., or $\frac{\text{Span}}{2}$ = Total run.

Note: The total run of the common rafter is always one-half of the span.

Unit of Run.—(Figure 4.) The unit of measurement, 1 ft., or 12 in., is the same for the roof as for any other part of the building or the lumber with which it is built, and it is by the use of this common unit of measurement that the framing square, a comparatively small instrument, can be employed in laying out the largest roof. The principle underlying all roof framing is that all run measurements are taken in a *level plane*. The level-plane distance is always spoken of as *run;* hence the unit of run, as before stated, is 1 ft., or 12 in.

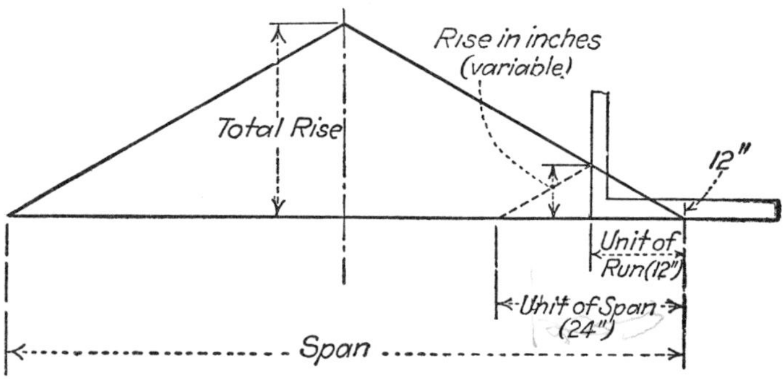

FIG. 4.

Unit of Span.—As the span is always twice the run, the unit of span is twice the unit of run. Therefore 2 times 12 in., or 24 in., is the unit of span.

Rise in Inches.—The rise in inches is the number of inches that any roof "rises" to every foot of run.

Pitch.[1]—"Pitch" is the term used to describe the amount of slope of the roof. It is equal to the total rise divided by the span. The total rise is the *vertical* distance that the ridge is above the plate line.

To illustrate:

The span of a building is 27 ft., the total rise is 9 ft. What is the pitch?

Solution: The total rise, 9 ft., divided by the span, 27 ft., equals the pitch; or 9 divided by 27 equals 9/27, or 1/3 pitch. $\frac{\text{Rise}}{\text{Span}} = \text{Pitch}.$

[1] See problem 2, Chap. VI, page 102.

Cut.—(Figure 5.) The "cut of a roof" is the rise in inches *and* the unit of run (12 in.).

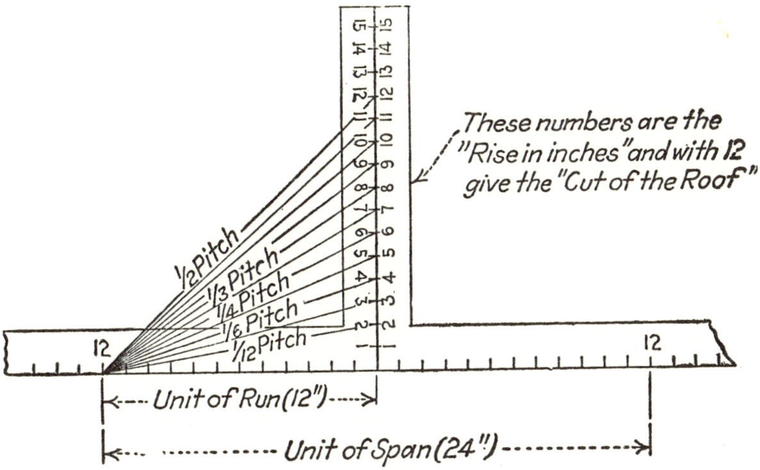

FIG. 5.

To illustrate:

The rise in inches is 8. Therefore, the cut of the roof is 8 and 12. The rise is 4 in. The cut is 4 and 12. The cut forms the basis of the layout process of any rafter.

Mathematics of "Cut."[1]—As stated under Total Rise, Pitch × Span = Total rise. Likewise the *pitch times the unit of span* (24 *in.*) will give the rise in inches.

To illustrate:

The pitch is ⅛. What is the cut?
Solution: Unit of span (24 in.) × ⅛ equals 3, or rise in inches. Pitch × Unit of span = Rise in inches. If rise in inches is 3, then the cut is "rise in inches" and unit of run, or 3 & 12.

Line Length.—(Figures 6 and 7.) The term "line length" as applied to roof framing is the hypotenuse of a triangle whose base is the total run and whose altitude is the total rise (see Figs. 2 and 3). This "line length" is just what its name implies. No thickness or width of stock is considered; it is simply a linear measurement.

[1] See problem 17, Chap. VI, page 117.

To find the line length, apply the mathematical principles that have been developed. The pitch of the roof times its span equals the total rise, and one-half its span equals the total run.

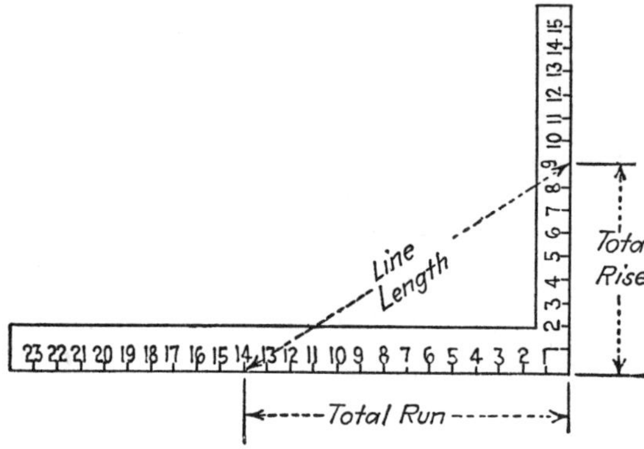

FIG. 6.

With a scale of 1 in. to the foot, or a $\frac{1}{12}$ scale, a framing square can very conveniently be used to represent two legs of the before-mentioned triangle by letting the blade be the run and the tongue of the square the rise. The hypothenuse of the triangle (the line that represents the rafter length) can be measured by another square, and by using the $\frac{1}{12}$ side of this square the length of the rafter can be determined to the exact inch, each division being equal to 1 in.

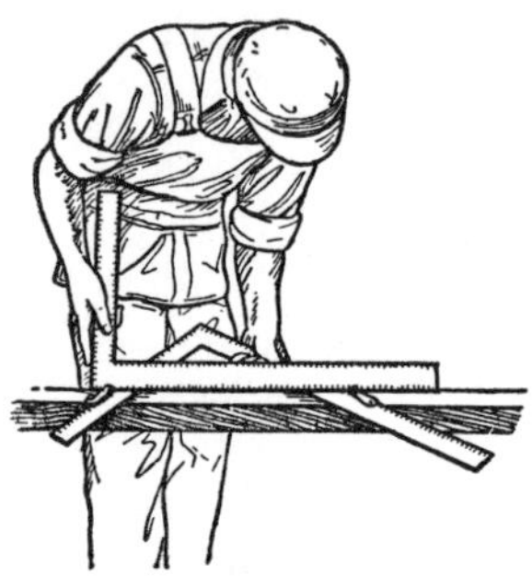

FIG. 7.

Note: Not all makes of squares are graduated on the back into twelfth divisions (maker's name is stamped on the face side). However, as this type of square is the best to use, it is the kind recommended.

If possible, always let the tongue of the square represent the rise cut and the blade represent the run cut. This is important because in a complicated cut it is sometimes hard to determine which part of the square to mark against. When used as above, if a rise cut is needed, mark on tongue; if a level cut is desired, mark on blade. The scale line length can be found by measuring across the square from a given total rise and total run. This is best done by using the 12th side of another square (see Fig. 7). Multiply the scale line length by 12 and the actual mathematical length is obtained. If the total rise and the total run cannot be represented on the square to the 1-in. scale, reduce to the ½-in. or ¼-in. scale to the foot, and proceed as before.

Square gages, as illustrated in Fig. 7 serve as a splendid aid in establishing different measurements on the square and also simplify the problem of marking the necessary different angles. Clamp the gages on the square to the required rise and run and place the square against the top edge of the rafter. The gages will keep the square to the correct angle while the layout work is being done.

Plumb and Level Lines.—(Figure 8.) These terms have reference to the direction of a line on a rafter and not any

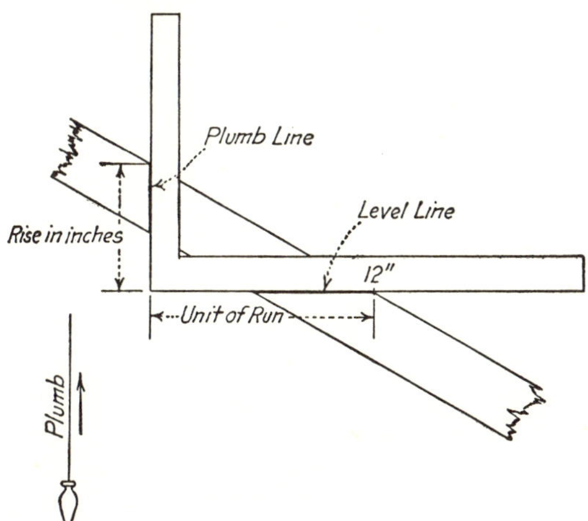

FIG. 8.

particular rafter cut. Any line that is vertical when the rafter is in its proper position is called a *plumb line,* and any line that is level when the rafter is in its proper position is called a *level line.*

CHAPTER II

THE GABLE ROOF

I. The Common Rafter:

- Layout process:
  - Line length method
- Step-off method
- Ridge cut
- Seat cut
- Overhang
- Tail cut
- Shortening

II. The Principle of Shortening

III. The Ridge Board

CHAPTER II

THE EQUAL PITCH GABLE ROOF

An equal pitch gable roof (Fig. 9) is a roof that has an equal pitch or slant in two opposite directions. It is formed by two plane surfaces intersecting at a ridge. This type of roof is the simplest roof to frame as it is composed of only two parts: namely, the common rafters and the ridge. A ridge piece is

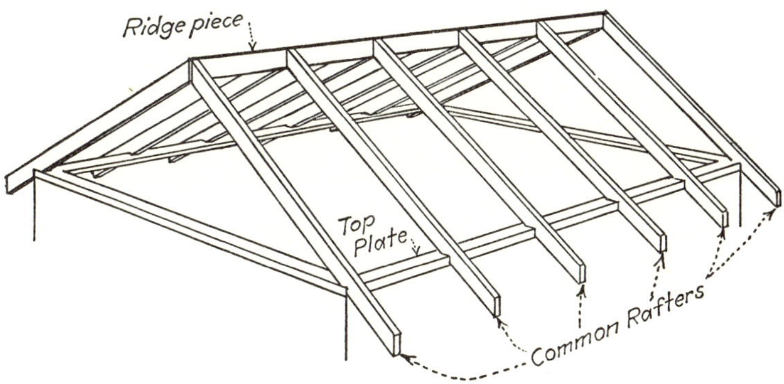

Fig. 9.

not absolutely necessary but its use insures a better form of construction and simplifies the job of "raising" the rafters.

THE COMMON RAFTER (FIGURE 10)

A common rafter is that member of a roof that extends at right angles from the plate line to the ridge. It is called a *common* rafter because it is common to all types of roofs and is used as the basis for the "layout" process of the other kinds of rafters in a roof.

The several cuts of the common rafter are the *ridge cut* at the top end of the rafter, the *seat cut* at the building line, and the *tail cut* at the exposed end. The distance between the ridge cut and the seat cut is called the *line length* (see p. 5, Chap. I). The part of the rafter that extends beyond the building line is called the *overhang*.

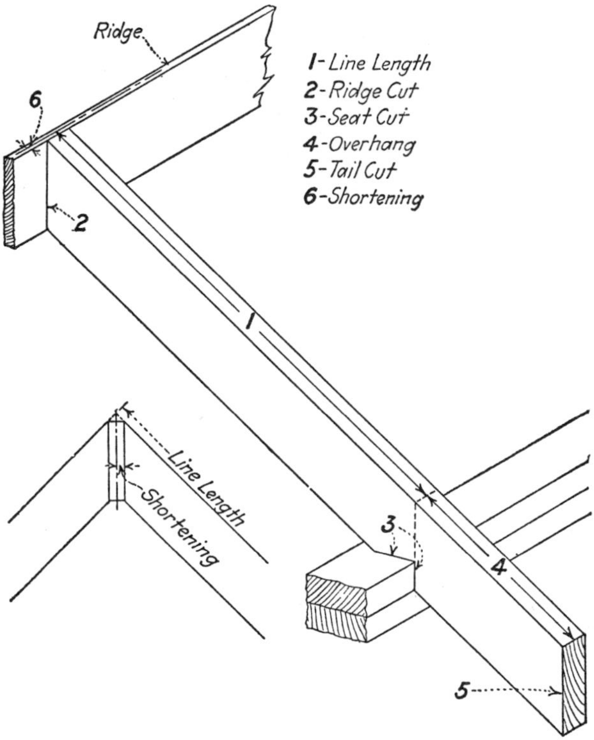

FIG. 10.

The Layout Process of a Common Rafter.—There are several methods of laying out the common rafter. Before any actual layout work can be done, however, the length of stock required for a given rafter must be determined. This is best done by the "line length" method, as illustrated in Fig. 6.

After the stock is selected, begin at the top, or ridge end, of the rafter and work toward the bottom. This permits of variation in the overhang distance, which is quite essential, as it is sometimes necessary and convenient to change the overhang distance to suit: (1) either the length of the rafter stock, or (2) the width of sheathing stock used on the overhang.

Some allowance will have to be made for the thickness of the ridge stock as the length of the rafter is considered to the center of the ridge. This allowance is called shortening (see Fig. 16).

Common Rafter Line Length.—(Figure 11.) There are two methods of finding the line length of a common rafter:

A. The scale method.
B. The stepping-off method.

The principles of method *A*, were described and illustrated on page 6, Chap. I. To apply this principle to the common

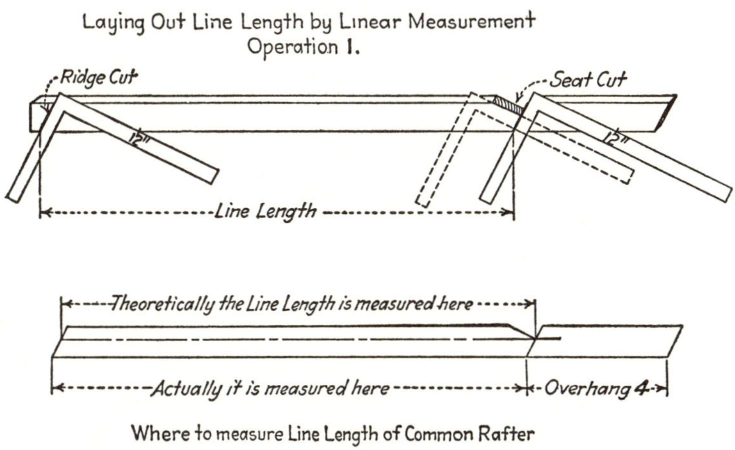

FIG. 11.

rafter, represent on the steel square the total rise and the total run to a $\frac{1}{12}$ scale. The diagonal distance across the square from gage to gage will be the scale line length of the common rafter (not including the overhang).

The scale length represents the net length of rafter stock required *before* the rafter is laid off, and is of great advantage as it enables one to determine the length of stock to use for any particular rafter. Overhang must be added, however.

As shown in Fig. 11, the length of a rafter can be measured on its top edge irrespective of the height of the seat cut. This is true because, whatever stock is left above the seat-cut line, the top end of the rafter will "raise" a similar amount. Therefore, irrespective of the width of the rafter stock, all measurements can be taken along the top edge of the rafter.

How to Use the Line Length.—(Figure 12.) Determine the line length of a given rafter. Also find the "cut" of the rafter and set gages on square to rise and unit of run. Hold square with the blade in the right hand and the tongue in the left, as illustrated in the drawing. This will bring the 12th side of the square up where it can be easily read, when desired.

FIG. 12.

Now place the square on the flat side of the rafter stock and at the extreme left end. Care should be taken that the crowned edge of the rafter stock is towards the operator, as that edge will be the top of the rafter when it is in its raised position. In framing, "crown" edges of stock are always placed up to allow for the natural sag due to its own weight.

Next mark a line at the tongue side of the square. This is the ridge-cut line and also a plumb line when the rafter is in its position on the building. Measure from this plumb line a distance equal to the line length of the rafter and mark another line. This will be the plumb line of the seat cut and locates that part of the rafter that "lines up" with the outside edge of the top plate.

The Principle of the Step-off Method.—(Figure 13.) The "stepping-off" method, *B*, is based on the unit of run (12 in.). With the gages on the steel square set to the cut of the roof (see Fig. 5), place the square in position on the top edge of the rafter stock. Next, draw a line at each outside edge of the tongue and blade of the square. The length of this line is the length of the common rafter for 1 *ft. of run.*

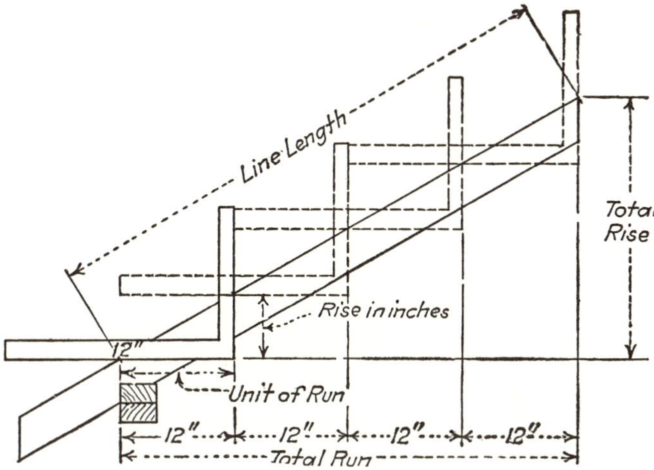

Fig. 13.

For 2 ft. of run the square should be moved to the second position. For 3 ft. of run, the square should be moved to the third position, etc. Stated as a rule, "move the square as many times as there are total feet of run, marking each position carefully so as to get accurate results."

Step-off Method as Applied to the Common Rafter.—(Figure 14.) To use the step-off method, proceed as follows.

Lay off and mark the ridge cut, as outlined on page 14, with the square in position *A* of the drawing, mark at both edges of the square as shown. Next shift the square to position *B* and mark. Continue this process until the square has been moved as many times as there are total feet of run. When the square has been moved the required number of times, mark a plumb cut line. This line will be parallel to the ridge cut.

For the fractional part of the unit of run the same principle can be applied. Place the square in position, but, instead of marking at the blade edge of the square, mark at *any one* of the inch marks, along the blade, depending upon the fractional part of a foot desired.

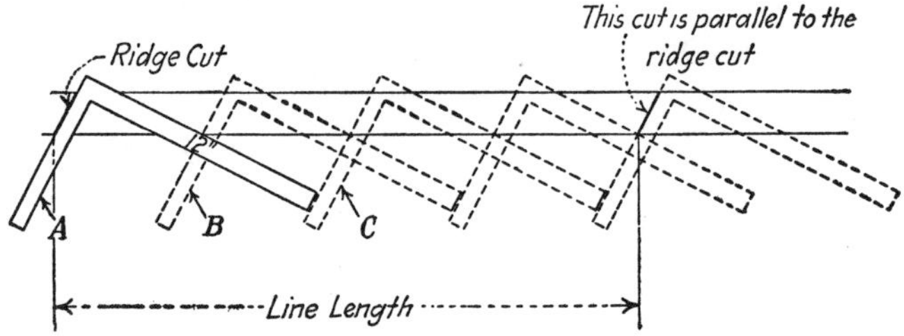

Lay-out by "Step-off" method

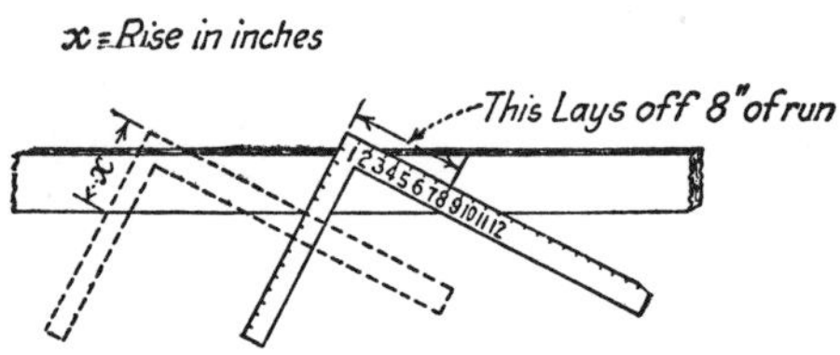

How to layout a fractional part of the "Unit of Run"

Fig. 14.

To illustrate:

For a roof that has a total run of 10 ft. and 8 in. step off for the 10-ft. run as usual. Next shift the square as shown and mark at the 8-in. mark. A plumb line at this point will then give 10 ft. and 8 in. of run.

The Ridge Cut of a Common Rafter.—(Figure 15.) The ridge cut of a common rafter is the cut at its top end and fits against either the ridge piece of the roof or, if no ridge is used, the opposite common rafter.

Where no ridge piece is used the plumb line, from which the length of the rafter is measured, becomes the ridge cutting line.

If the roof has a ridge, allowance must be made for the thickness of the ridge stock. This is explained below. In either case, set the gages on the steel square to the "cut" of the roof and mark as illustrated in Fig. 14.

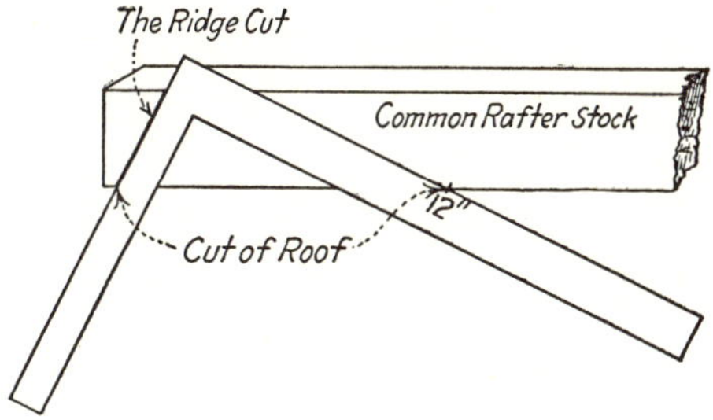

Fig. 15.

Shortening of the Common Rafter for the Ridge Piece.—(Figure 16.) As the ridge piece is placed at the intersection of

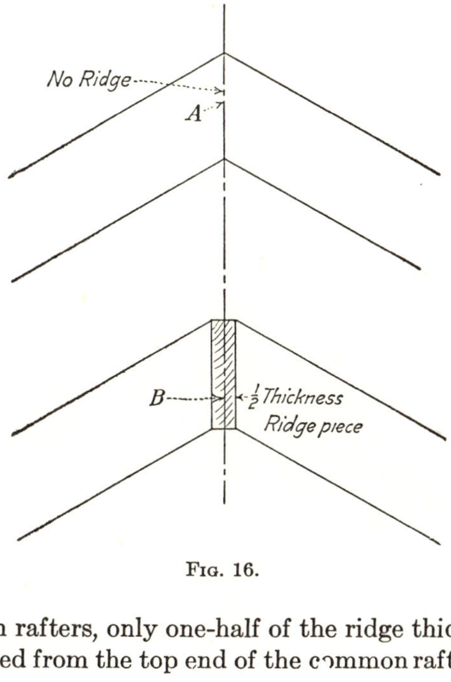

Fig. 16.

two common rafters, only one-half of the ridge thickness needs to be deducted from the top end of the common rafter. Therefore, measure the ridge piece and divide the thickness by two. Next measure this distance at *right angles* from the plumb line, *A*, at the top end of the rafter stock, and draw another plumb line *B*. This is the line on which to cut the rafter.

Note that the slant or pitch of the roof does not affect the shortening, if the shortening measurements are always taken at *right* angles to the plumb cut. Too much emphasis cannot

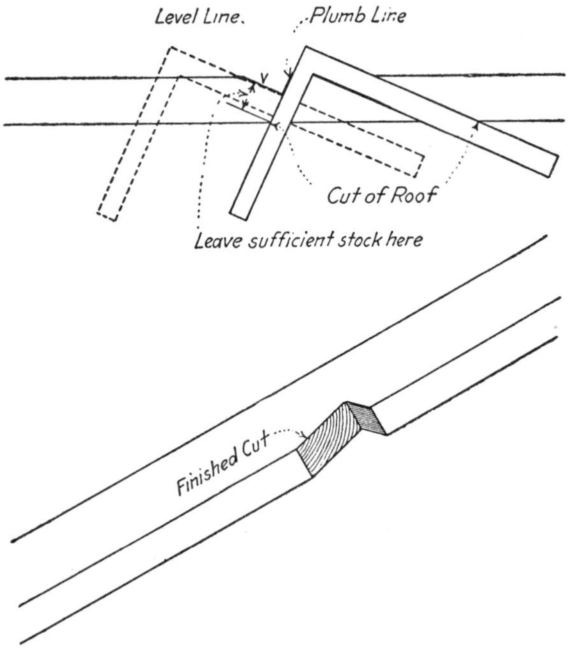

FIG. 17.

be placed on this fact, as a fundamental principle of roof framing is that measurements taken *on the level* will always be the same for any pitch roof.

Seat Cut of the Common Rafter.—(Figure 17.) The seat cut of the common rafter is the cut that permits the rafter to fit or rest on the top plate of the building. It is composed of both plumb and level lines.

The plumb line is laid off to the cut of the roof. (If the line length method has been used, it is already marked on the rafter.)

Note that the level line is at right angles to the plumb line.

On the plumb line, locate a point that will leave sufficient stock above the plate line to insure the strength necessary for the overhanging part of the rafter. Using the steel square, mark a level line at the point so located. Cutting on these lines as illustrated forms the seat cut.

Overhang of the Common Rafter.—(Figure 18.) The overhang of the common rafter is that portion of it which extends

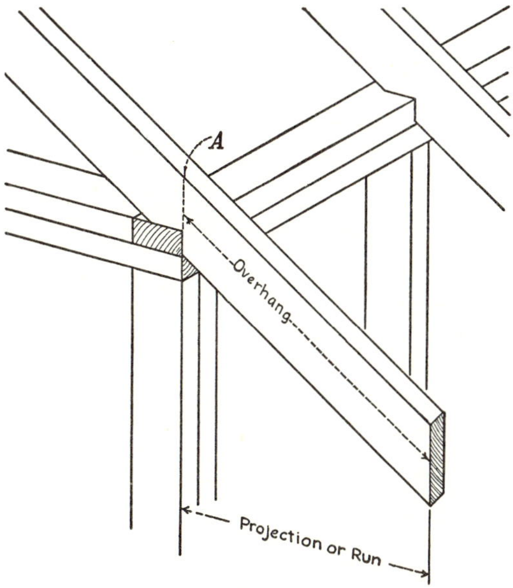

Fig. 18.

beyond the building line. The distance that the rafter extends is determined by several factors: namely, the design of the building, the length of the common rafter stock, or the width of the sheathing material used. In any case, the point from which to measure is *always* the plumb line of the seat cut (see *A* in Fig. 18).

Either measure on the top edge of the rafter, or, if projection (run) measurement is given, use the step-off method. Therefore, determine the overhang distance by one of the following methods:

1. Multiply the width of one cornice board (usually T & G stock) by the number of boards desired.

2. Measure the "run" of the cornice on the plan, and step off the required length, or,

3. Measure the actual scale length on the elevation sheet of the plans or on the cornice detail.

Tail Cut of a Common Rafter.—(Figure 19.) Little can be said concerning the shape of the cut at the exposed end of a

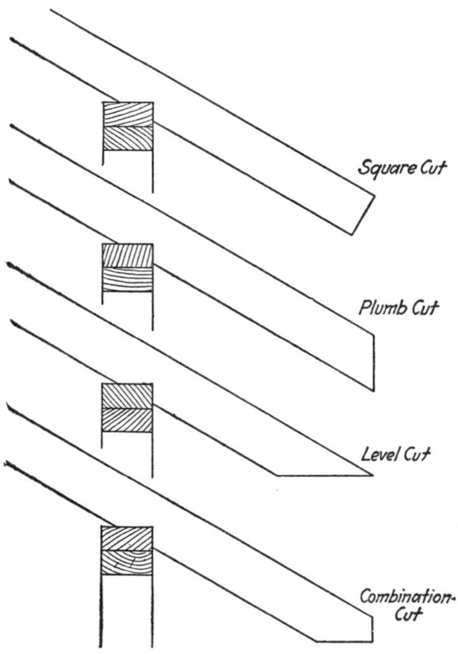

FIG. 19.

common rafter, as the possibility of different designs is practically unlimited. Figure 19 illustrates a few typical cuts that are commonly used. Generally speaking, they are a combination of the level and plumb cut and as such can be laid out by the use of the steel square in the same way as any plumb or level cut.

LAYOUT FOR THE SHORTENING OF THE COMMON RAFTER (FIGURE 20)

Figure 20 gives a side view of the ridge end of a common rafter with the plumb lines drawn on the stock so as to allow for the shortening which has been discussed in the preceding pages and illustrated in Fig 16. Note that distance "B" is measured *in the level plane* which is at right angles to the plumb line. Allowance for shortening is *always* measured in this direction, *regardless* of the slant or pitch of the roof.

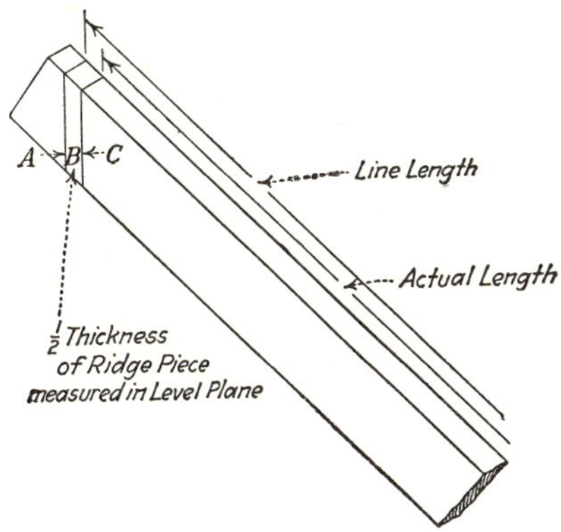

FIG. 20.

For the common rafter, *the amount to shorten* is one half of the square thickness of the ridge stock because this rafter, is always at a 90 degree angle to the ridge.

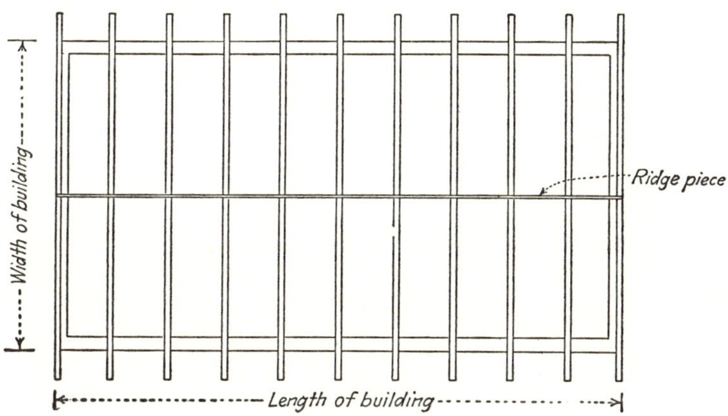

FIG. 21.

THE RIDGE

Length of Ridge Piece of a Gable Roof.—(Figure 21.) The ridge piece of a gable roof is always the same length as that part of the building to which it is parallel.

CHAPTER III

THE EQUAL PITCH HIP ROOF

I. The Hip Rafter:

- Methods of construction
- The run position
- The unit of run
- Cut of the hip
- Line length by:
  - The slant triangle method
  - The step-off method
- Ridge cut when framed to:
  - The ridge
  - The common rafter
- Shortening when framed to:
  - The ridge
  - The common rafter
- Seat cut
- Drop or backing
- Overhang
- Tail cut

II. The Cheek Cut:

- Kinds of cheek cuts
- Principle of the cheek cut
- The single cheek cut layout process
- The double cheek cut layout process

III. The Hip Jack Rafter:

- Cuts
- Layout process
- Line-length methods
- Cheek cut
- Seat cut
- Tail cut
- Overhang

IV. The Ridge:

- Length when hip is framed to the ridge
- Length when hip is framed to the common rafter

CHAPTER III

THE EQUAL PITCH HIP ROOF

An equal pitch hip roof differs from a gable roof in that its slant or slope is in four directions instead of two (see Fig. 22).

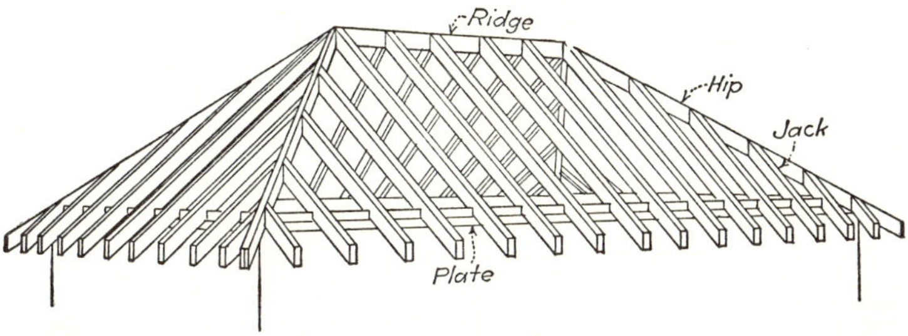

FIG. 22.

A hip roof, like a gable roof, has common rafters and a ridge, and, in addition, hip rafters and hip jack rafters. As the common rafters have been fully discussed in Chap. II, no further description need be given.

THE HIP RAFTER

The angle formed by the intersection of the sloping surfaces of a hip roof makes a rafter necessary at that place. This rafter is called the *hip rafter*. For a full-hipped roof, four hip rafters are required.

Methods of Construction.—There are *two* common methods of framing a hip roof: method *A*, illustrated in Fig. 23, which shows the top end of the hip rafter framed to the ridge piece; and method *B*, which shows the top end of the hip rafter framed to the intersection of three common rafters. The latter might be termed the "tripod" method, as the three common rafters shown in the drawing form a tripod when in their proper position in the roof.

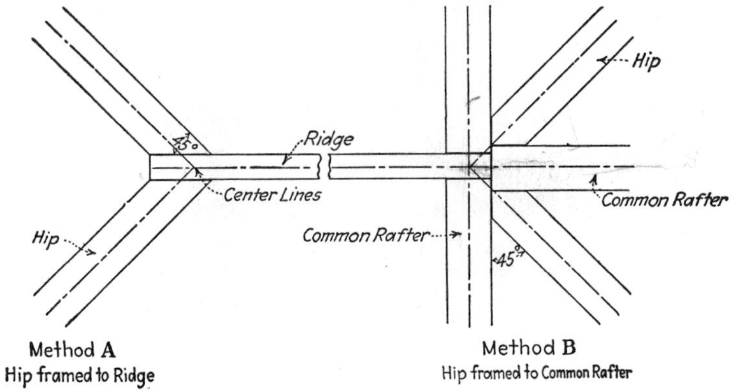
45°
Ridge
Center Lines
Hip
Hip
Common Rafter
Common Rafter
45°
Method A
Hip framed to Ridge
Method B
Hip framed to Common Rafter

FIG. 23.

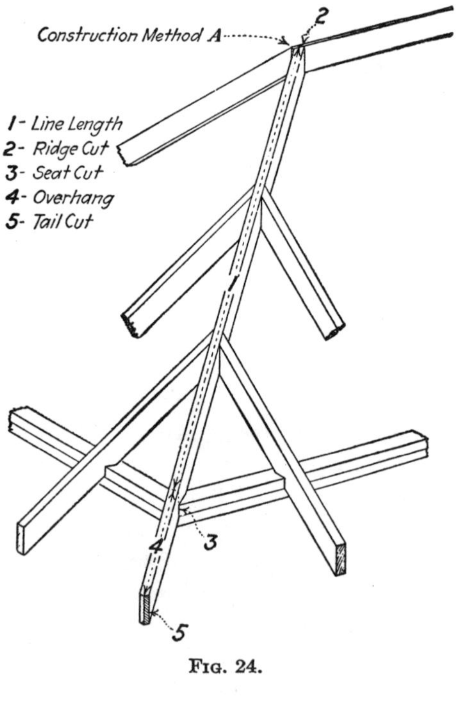
Construction Method A
2
1- Line Length
2- Ridge Cut
3- Seat Cut
4- Overhang
5- Tail Cut
1
3
4
5

FIG. 24.

Operations in Framing a Hip Rafter.—(Figures 24 and 25.) The operations in laying out a hip rafter are as follows:

1. Find the line length of the hip rafter.
2. Lay out the ridge cut.
3. Lay out the seat cut.
4. Find the amount of overhang.
5. Lay out the tail cut.

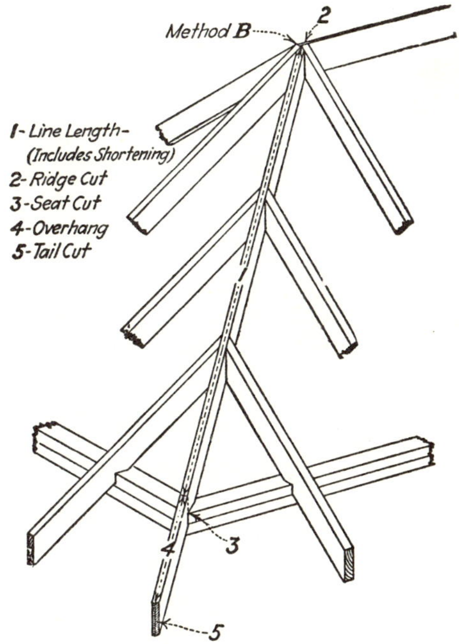

Fig. 25.

These are illustrated in the drawings. The layout process is identical for either *A* or *B* construction.

The Cuts and Terms of a Hip Rafter.—(Figure 24.) The cuts of the hip rafter are similar in name to the cuts of the common rafter. Each hip will have a *ridge cut* at the top end, a *seat cut* at the building line, and a *tail cut* at the exposed end. The distance between the ridge cut and the seat cut is called the *line length*.

The hip rafter cuts, however, are quite different in shape from those of the common rafter. Each of these cuts is discussed separately. As explained on page 17, the overhang of the hip rafter is the same as that of the common rafter and refers to that part of any rafter which extends beyond the building.

Figures 24 and 25 show the hip rafter in place with the alternative constructions, *viz.*, framed to the ridge (Fig. 24), and framed to the common rafters (Fig. 25).

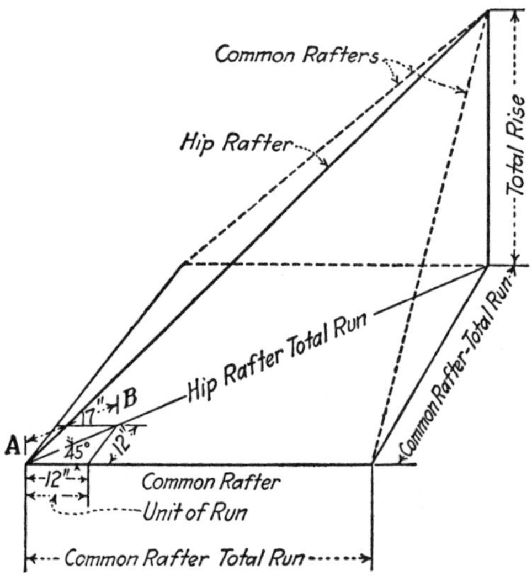

A-B Hip Unit of Run = 16.97" or 17"

FIG. 26.

The Layout Process.—As in the case of the common rafter, the best method of procedure when laying off the hip is to begin at the top and work toward the bottom end. This permits of a variation of the overhang distance. Before a hip can be "laid out," however, some fundamental mathematical principles must be discussed in order to develop a basis upon which to work.

Run Position of the Hip Rafter.—(Figure 26.) A new principle in roof framing arises in laying out the hip roof, due to the fact that the run of the hip rafter is *not* at right angles to the plate line of the building, but at an angle of 45 deg. This position forms the basis on which the shape of the hip rafter cut is developed.

Unit of Run.—(Figure 26.) The "unit of run" of the hip rafter is based on the unit of run of the common rafter (12 in.). For every foot of common rafter run, the hip has for its unit of run the diagonal of 12 and 12, or 16.97 in., which for practical purposes is always considered 17 in.

Cut of Hip.—(Figure 27.) The hip rafter will "rise" no higher than the common rafter. Therefore, the cut of the hip

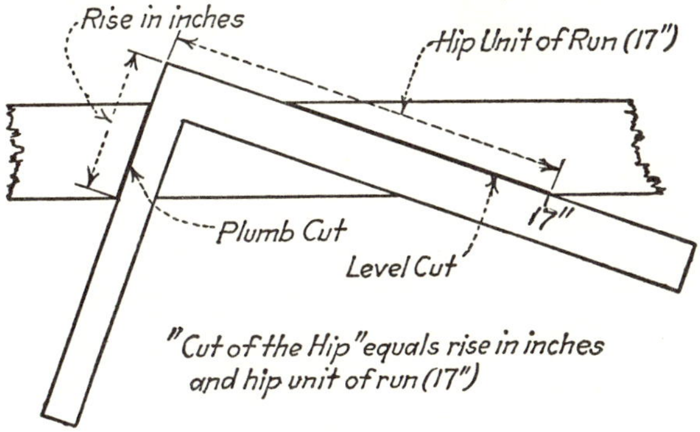

Fig. 27.

must be the rise in inches of the common rafter and the hip unit of run (17 in.).

Line Length of the Hip by the "Slant Triangle" Method.—(Figure 28.) The figure illustrates the "slant triangle" method of determining the line length of the hip rafter. This slant triangle is the triangle in the roof that is formed by the line length of the common rafter and one-half the width of the building. This is the simplest method for determining the hip line length, because the common rafter line length measurement and the span measurement are usually known, or can easily be determined.

Therefore, set the gages on the square so that they represent common-rafter line length and one-half the span. The diagonal distance from gage to gage is the scale line length of the hip rafters. The proper length of the hip rafter stock can now be selected, allowing sufficient stock for the overhang, if there is any. This is explained on page 36.

Note that if the base and common rafter lines were thought of as being in the width instead of the length of the building,

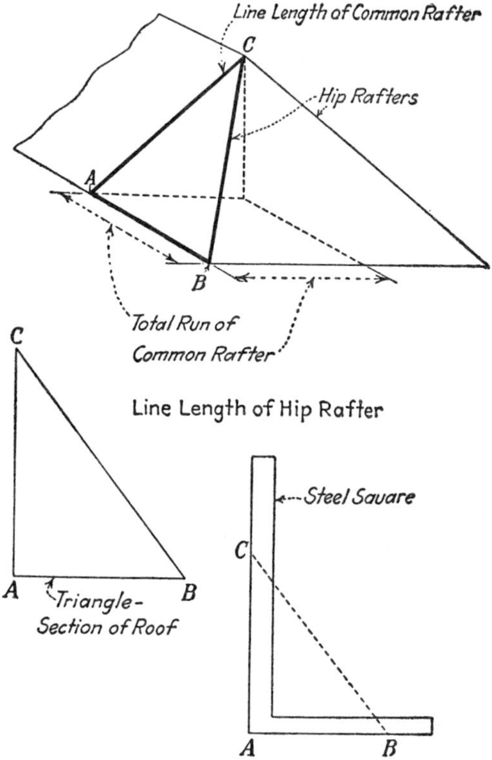

FIG. 28.

the triangle would still be identical in shape and size to the one shown.

How to Use Line Length.—To lay off a hip rafter after the line length is known, mark the ridge cut as usual, and (using a rule or tape), on the top edge of the rafter stock measure the required distance and mark the plumb line of the seat cut.

Line Length of the Hip Rafter by the Step-off Method.—(Figure 29.) The basis of the step-off method for any rafter is the *unit of run.* As stated previously, the unit of run for the common rafter is 12 in. and the unit of run for the hip 17 in., as illustrated in the drawing.

In stepping off, either a common or a hip rafter for a given roof, the number of steps taken is the same. The size of the steps however differs. That is, for every 12-in. run of the common rafter, the hip has a step (or run) of 17 in. Therefore,

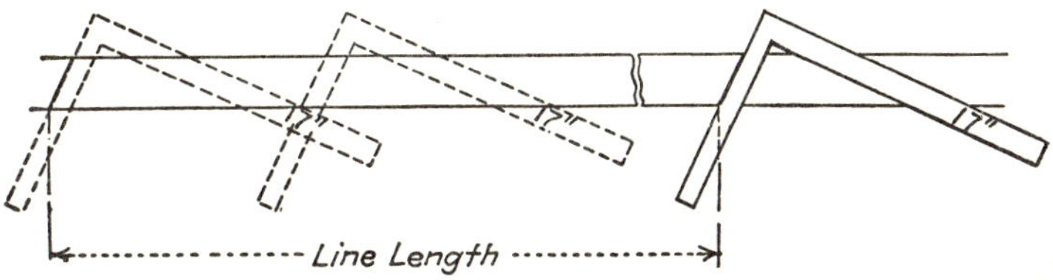

Line Length by the Step-off Method

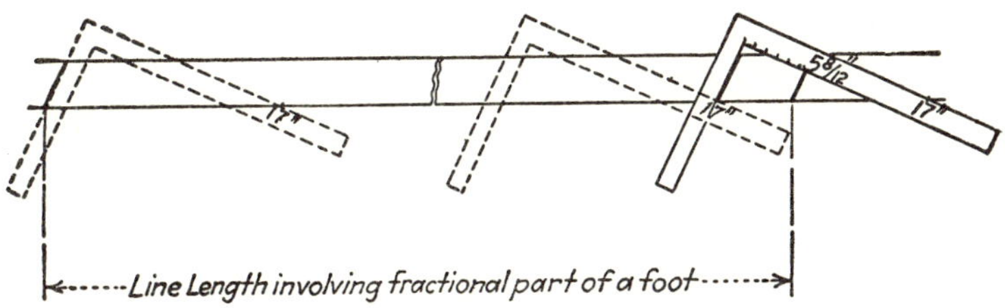

Method of Finding Line Length
When The Fractional Part of a Foot is Involved

FIG. 29.

set the gages to the "cut" of the hip rafter, *rise in inches* and 17, as shown in Fig. 29, and step off as many times as for the common rafter of the same roof. Fractional parts of the hip unit of run are marked off similarly to the fractional parts of the common rafter unit of run (see p. 14). This time, however the fractional part is *applied to the* 17 *in.*

To illustrate:

To step off a hip rafter for a building that has a span of 28 ft. and 8 in. and a ⅓ pitch roof: The total run is one-half of 28 ft. and 8 in., or 14 ft. and 4 in. Set the gages at rise in inches (8 in.) and 17, and step off 14 times. The 4 in. is one-third of the foot unit of run. Therefore, one-third of the 17 in. must be measured, which on the square is $5\frac{8}{12}$ in. Locate this point of the square, mark on the rafter stock, and draw a plumb line.

Note: As $5\frac{8}{12}$ will not come opposite the rafter stock, move the square over the width of the tongue ($1\frac{1}{2}$ in.) and mark on the inside of the blade (see figure).

Shape of the Ridge Cut of a Hip Rafter.—(Figure 30.) The ridge cut of a hip rafter is not a square but a slanting plumb

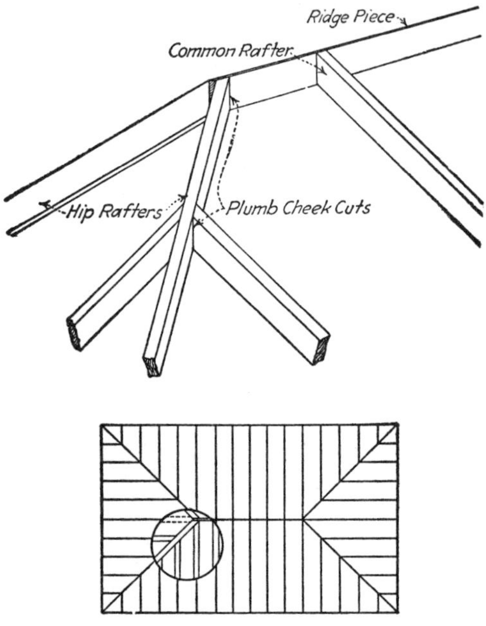

FIG. 30.

cut, and is known as a *cheek cut.* (Jack rafters and valley rafters also require this cut.)

Kinds of Cheek Cuts.—There are two kinds of cheek cuts, namely, the *single cheek cut* and the *double cheek cut,* as illustrated in Fig. 31. The layout process is the same for both. The direction of the angle of cheek cut is determined by the position of the rafter in the roof.

Layout Principle of the Cheek Cut.—(Figure 32.) On account of the many variations possible in the slant or pitch of any roof,

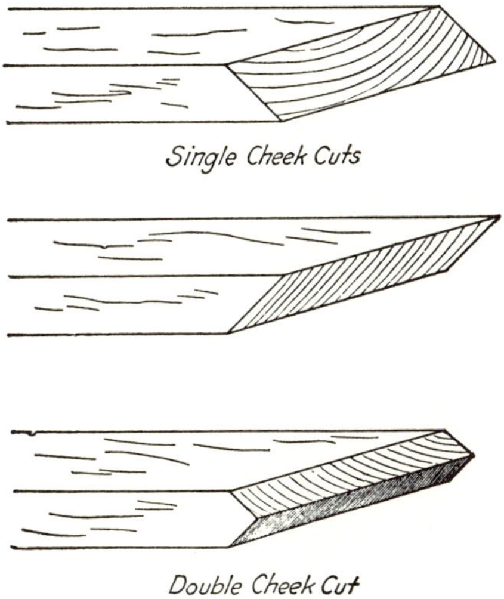
Single Cheek Cuts
Double Cheek Cut

FIG. 31.

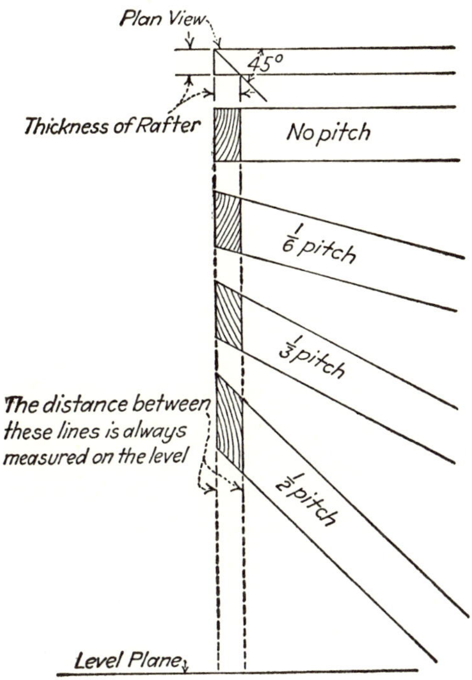
Plan View
45°
Thickness of Rafter
No pitch
1/6 pitch
1/3 pitch
1/2 pitch
The distance between these lines is always measured on the level
Level Plane

FIG. 32.

no measurements can be taken on the slope of the roof when laying out cheek cuts that would give the same result for different pitches. There is, however, a method of measuring that can be applied to any roof, irrespective of the pitch, that will *always* give the same results.

That part of the building which is common to all roofs is the *level plane* over which they are constructed, and any measurements on a rafter taken in the level plane, no matter what the slant of the roof is, will always be the same. As a level line is always at *right angles* to the plumb line, measuring at right angles to a plumb line is measuring in the level plane.

How to Lay Out a Single Cheek Cut.—(Figure 33.) The principle involved in the problem of laying out a cheek cut is

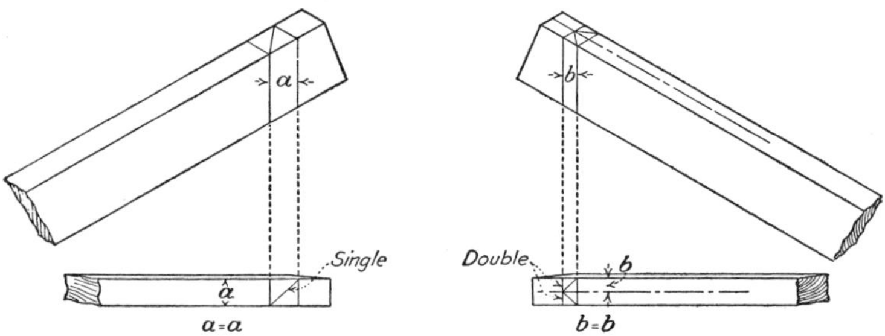

FIG. 33.

based on what actually takes place in the *plan view* intersection of any two roof members that are at an angle of 45 deg. to each other.

As shown in the drawing, when laying out a cheek cut on a rafter, the only method of procedure, as before stated, that will give the same result, irrespective of the pitch of the roof, is to measure in the *level plane*. Therefore, proceed as follows: Measure at *right angles* to the plumb cut line a distance equal to the thickness of the stock on which the cheek cut is being laid out. "Carry" both plumb lines across the top edge of the rafter and connect the diagonal corners. The direction of the diagonal line is determined by the position of the rafter in the roof.

How to Lay Out a Double Cheek Cut.—(Figure 33.) The double cheek cut is laid out in a manner similar to that for the single cheek cut. The measurement at right angles to the plumb cut line, however, is equal to only *one-half* the thickness of the stock on which the double cheek cut has been marked.

Shortening of Hip Rafters.—(Figure 34.) In "shortening" the hip rafter, allowance must be made for the material to

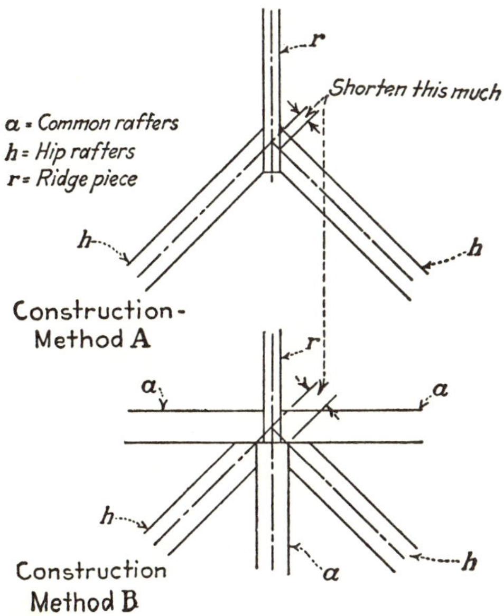

Fig. 34.

which the rafter is framed. In method *A* (Fig. 34), the ridge piece is involved, while in method *B* the common rafter is involved. In either case the shortening process is best done before the cheek cut is laid out.

Shortening for Method A.—As the hip is always at an angle of 45 deg. to the ridge piece, (in plan view), one-half of the 45-deg. thickness of the ridge piece is the amount to shorten.

Shortening for Method B.—In this case it will be one-half of the 45 deg. thickness of the common rafter stock.

How to Lay Out a 45-deg. Line.—(Figure 35.) To mark a 45-deg. line, use equal numbers on the square as indicated in the drawing. Mark at either tongue or blade.

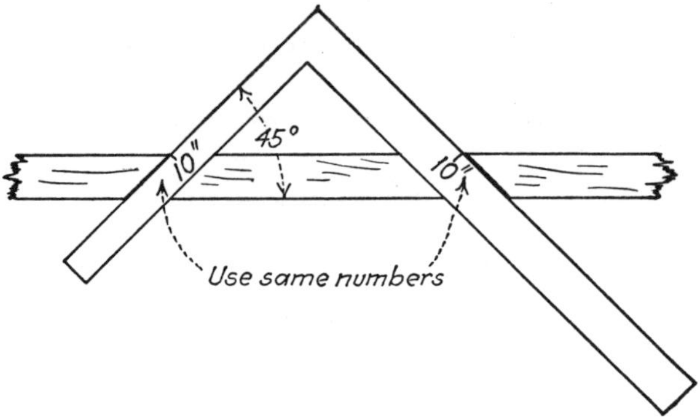

FIG. 35.

How to Shorten.—Mark a plumb measuring line at the top end of the hip stock, then measure at *right angles* to this line, the required amount which, as stated, depends upon the type of construction used. The important thing is to be sure that these measurements are taken at *right angles to the plumb cut.*

How to Lay Out a Ridge Cut of a Hip Rafter Using Method *A* Construction.—(Figure 36.) To lay out the ridge cut of a hip rafter that is framed to a ridge piece, using construction method *A*, proceed as follows:

1. Find the cut of the rafter and mark a plumb line near the left end of the rafter stock.
2. From this line, measure a distance equal to one half the 45-deg. thickness of the ridge piece and draw another plumb line.
3. From this second line measure a distance equal to one-half the thickness of the rafter stock, and draw a third line on both sides of the rafter.
4. Carry all three lines across the top edge of the rafter stock.
5. Draw a center line on the top edge.
6. Draw a cheek cut line as shown. The position of the rafter will determine in which direction to draw it.

Cut the rafter on this cheek cut line and the third plumb line, as shown in the drawing.

How to Lay Out the Ridge Cut of a Hip Rafter Using Construction Method *B*.—(Figure 37.) In laying out the ridge

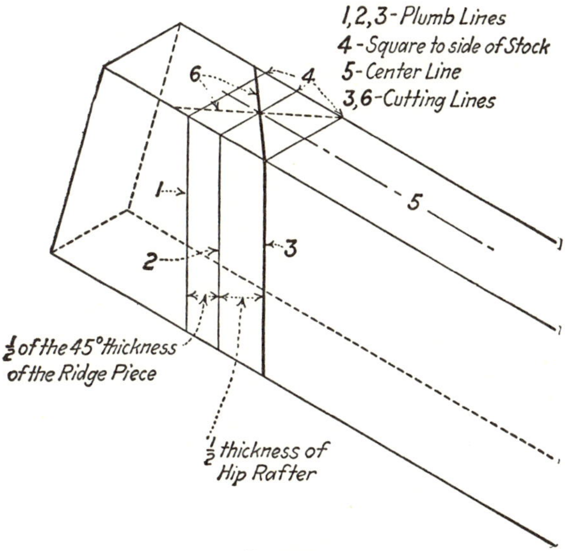
1,2,3-Plumb Lines
4-Square to side of Stock
5-Center Line
3,6-Cutting Lines
6
4
1
5
2
3
½ of the 45° thickness of the Ridge Piece
½ thickness of Hip Rafter

FIG. 36.

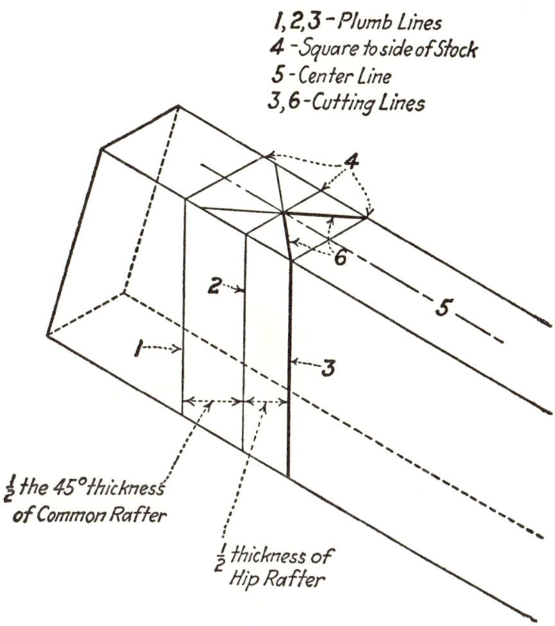
1,2,3-Plumb Lines
4-Square to side of Stock
5-Center Line
3,6-Cutting Lines
4
6
2
5
1
3
½ the 45° thickness of Common Rafter
½ thickness of Hip Rafter

FIG. 37.

cut of a hip rafter that is to be framed to a common rafter, the procedure in the main is the same as if the rafter were framed to a ridge. The measurement for shortening, however, is different, as the cut is a double instead of a single cheek cut. Therefore,

1. Find the cut of the rafter and draw a plumb line near the left end of the rafter stock.

2. From this line measure a distance equal to one-half the 45-deg. thickness of the common rafter stock and draw another line.

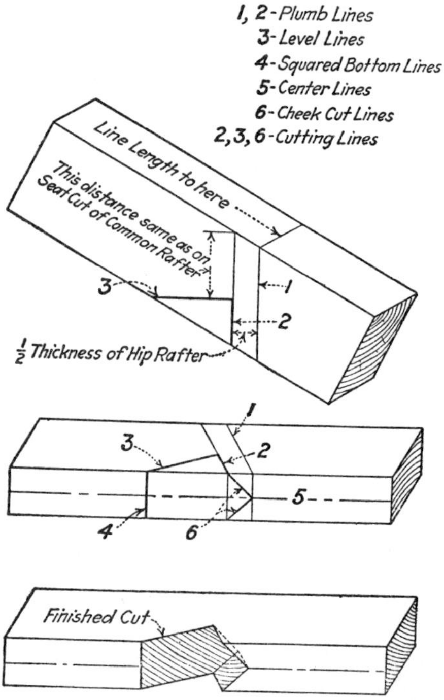

Fig. 38.

3. From this second line measure a distance equal to one-half the thickness of the rafter stock and draw a third line.

4. Carry all three lines across the top edge of the rafter stock.

5. Draw a center line on the top edge of the stock.

6. Draw cheek cut lines as shown. The rafter should be cut on the plumb line and the cheek cut lines.

How to Lay Out the Seat Cut of a Hip Rafter.—(Figure 38.) The seat cut of a hip rafter is a combination of level, plumb and cheek cut lines. The drawing illustrates the layout process and shows the shape.

In measuring the line length of the hip, a plumb line will have been marked on the rafter that "lines up" with the outside edge of the building. This line is also the plumb line of the seat cut and the measuring point for it.

The following steps will explain the order of procedure in laying out the seat cut:

1. Measure from the plumb line (1) *toward the top end* of the rafter a distance equal to one-half the thickness of the hip rafter stock, and draw another line (2).

2. Measure on line (2), which is the part of the rafter that touches the building, a distance equal to the amount of stock that was left above the common rafter seat cut.

3. Draw a level line at this point (3).

4. "Carry" these lines across the bottom of the rafter stock as at (4).

5. Locate a center line at the bottom of the stock as at (5).

6. Draw diagonal lines as shown from the center point on the line to the edge of line (2).

The rafter should be cut on lines (2), (3), (4), and (6).

Dropping or Backing the Hip Rafter.—(Figure 39.) When the sheathing stock is placed on the top edge of the hip rafter after that rafter is nailed in its proper position, it will be found that the sheathing strikes the outside top corners of the hip and leaves a space above the center of the hip. As all measurements in rafter layout work are always considered as being taken on the center line of the rafter, some allowance must be made when framing the hip rafter, since measuring the height of the seat cut on the line length line will bring the hip rafter *above* the plane of the roof.

There are two ways of taking care of this difficulty. Either "drop" the rafter by making the level part of the seat cut *higher*, thus letting the rafter drop, or else "back" the top edge of the hip by planing it on a bevel, the high part of the bevel being in the center of the hip.

How to Find How Much to Drop.—Dropping the hip is much the easier process of the two and can be done automatically when the seat cut is being laid out. Instead of measuring the height of the seat cut on the line length line, *measure it on the cheek cut line.* This cheek cut line represents that part of the hip that "lines up" with the outside of the top plate. This will "raise" the seat cut and "drop" the hip.

How to Find How Much to Back.—The amount of backing is

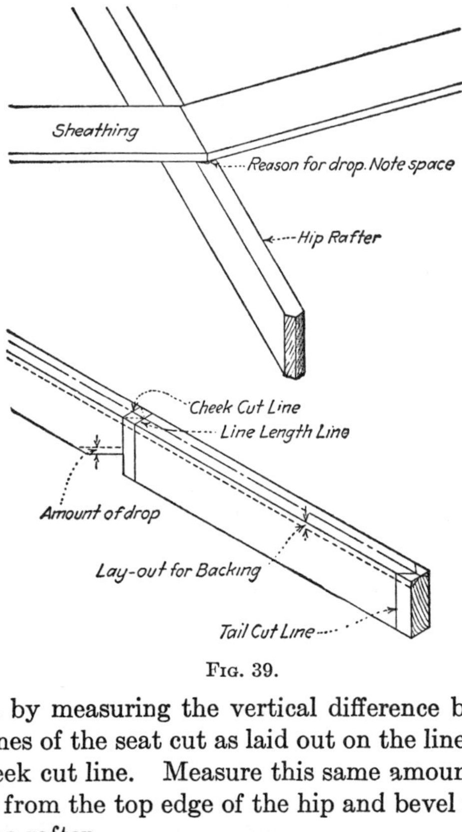

FIG. 39.

determined by measuring the vertical difference between the two level lines of the seat cut as laid out on the line length line and the cheek cut line. Measure this same amount in a *vertical position* from the top edge of the hip and bevel it off to the center of the rafter.

Overhang of the Hip Rafter.—(Figure 40.) Finding the overhang of the hip rafter presents a problem that is separate and apart from that of determining its line length.

To find the exact overhang length, the "*run*" *of the cornice* must first be known because the "cornice run" of the hip is the 45-deg. diagonal distance of the cornice run of the common rafter as shown in the drawing.

How to Find and Lay Out the Cornice Run of Hip Rafters.— Find the run of the cornice by stepping off the tail cut of the common rafter. After the run is known, find the 45-deg. run of the hip overhang by setting each gage on square to the cornice run and measure the bridge distance. If the amount of the cornice run extends beyond the limits of the square, use ½ scale and proceed as before. Using the cut of the hip (rise in inches and 17 in.) step off the cornice run of the hip, starting from the line length point of the seat cut.

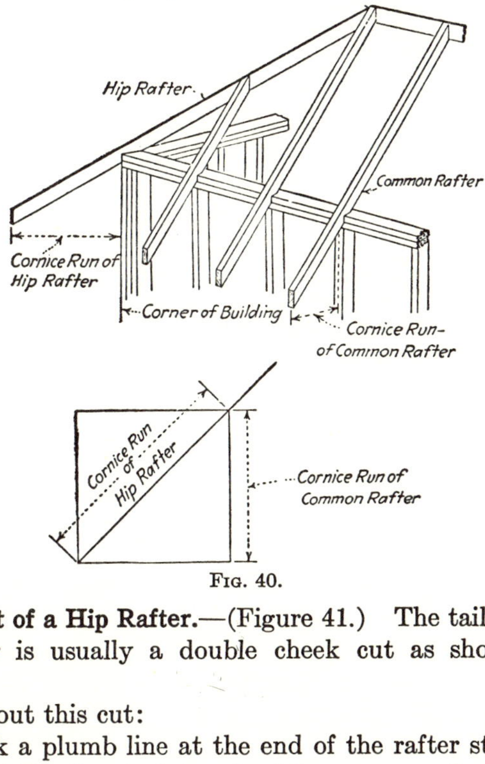

Fig. 40.

Tail Cut of a Hip Rafter.—(Figure 41.) The tail cut of the hip rafter is usually a double cheek cut as shown in the drawing.

To lay out this cut:

1. Mark a plumb line at the end of the rafter stock. The location of this line is determined by amount of overhang as described on page 36.

2. Measure from this plumb line *toward* the top end of the rafter, a distance equal to one-half the thickness of the rafter stock, and draw another plumb line (2). Note that this measuring is done *on the level.*

3. Draw a center line on the top edge of the rafter stock as at (3).

4. Draw angle lines as shown from center of line (3) to edge of line (2). Cut the rafter on the angle lines and line (2).

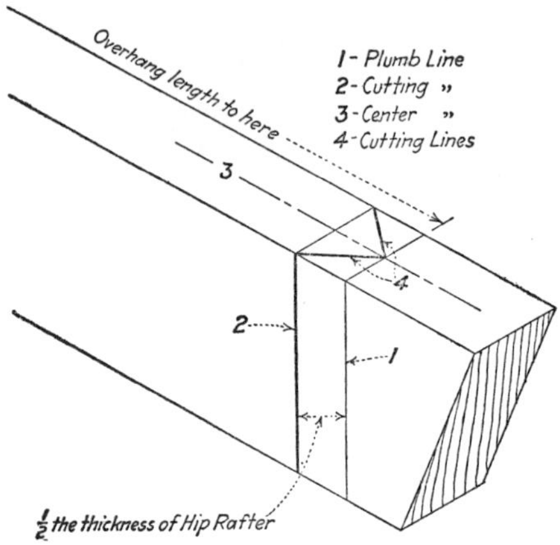

Fig. 41.

THE HIP JACK RAFTER

A hip jack rafter is in a sense a shortened common rafter as its lower end is framed identically with the common rafter (see Fig. 42).

Hip Jack Rafter Cuts.—The several cuts of a hip jack are the *cheek cut* at the top end of the rafter; the *seat cut* at the building line, which is exactly the same as the common rafter seat cut; and the *tail cut* at the exposed end of the rafter, which is also identical with the common rafter tail cut.

The *line length* of the jack is the mathematical distance between the seat cut and the cheek cut, *before* the shortening is allowed for. The actual length is the length of the rafter after it is cut and allowance has been made for the thickness of the hip rafter.

The Layout Process of a Hip Jack Rafter.—(Figure 42.) The general procedure in the layout process of the hip jack rafter is as follows:

1. Determine the line length of the hip jack.

2. Lay out the cheek cut (including the allowance for shortening).
3. Lay out the seat cut.
4. Determine the run and lay out the overhang.
5. Lay out the tail cut.

Note: Steps 3, 4, and 5 can be eliminated by using the common rafter pattern as this part of the jack is identical with that part of the common rafter.

Line Length Methods.—It may appear at first glance as though too many processes are given by which the length of a hip jack rafter is determined. There are really only two methods described, namely the "step-off" method and the "line length" method, but there are several ways of doing each

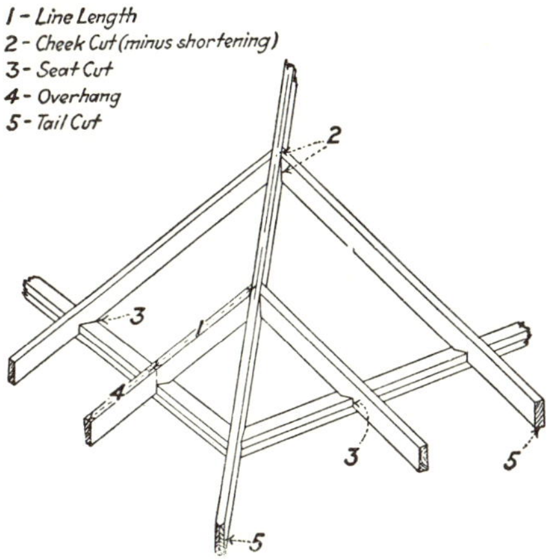

FIG. 42.

and experience has proved that each has its place, depending upon the circumstances involved; also what may prove an easy method for one man to understand may be a hard one for another. For example, for a complicated roof the method which requires a *roof layout* might prove the best and most economical and again, where only a few jack rafters are involved, the method based on *location* would undoubtedly be the quickest. Each process has been given with the intention that the craftsman should select the one that will be the quickest and best for the roof under consideration.

Line Length of a Jack Rafter by the Step-off Method Based on a Roof Layout.—(Figure 43.) The plan view of any roof irrespective of its pitch can be drawn only one way and would look as it appears to the eye when seen from some distance directly above the roof. That is to say, a roof might have any one of a dozen different pitches and yet the plan view of all would look exactly alike. This illustration will help to explain a very simple method by which the length of any rafter in a roof may be found.

Figure 43 represents a plan view or *roof layout* of a hip roof. *Every line that represents a rafter is the run of that rafter.*

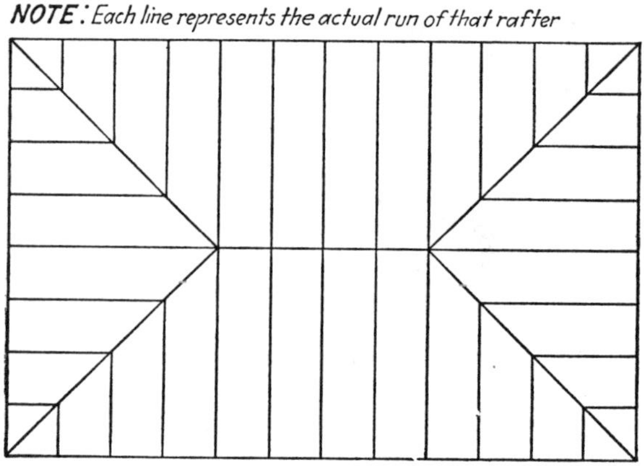

FIG. 43.

Therefore, to find the run of a hip jack rafter, draw a plan view of the roof involved, letting 1 in. represent 1 ft. This can be done on any smooth board, or even a wide shingle if the building is not too large.

Now measure any line that represents a hip jack and the length of the line will be the scale *run* of that jack.

To illustrate:

Assuming that the line measured 8 in., the rafter would have 8 ft. of run. If the line were 6¼ in. long, the rafter would have a corresponding run of 6 ft. 3 in.

Knowing the total run of a rafter, use the step-off method as was done for the common rafter, which is explained on page 13 of Chap. II. This is a good method to use if many rafters of different lengths are required.

Length of a Hip Jack by the Step-off Method Based on Location.—(Figure 44.) As explained in the previous paragraph, the run of a jack can be determined by the graphic method. Quite often, however, this is not a convenient method to use as drawing facilities are not available on the job and, even if they were, the job in hand may not warrant the time required to make such a roof layout.

Figure 44 illustrates a method of finding the total run of a hip jack rafter which is based on the *location* of the hip jack rafter in the roof.

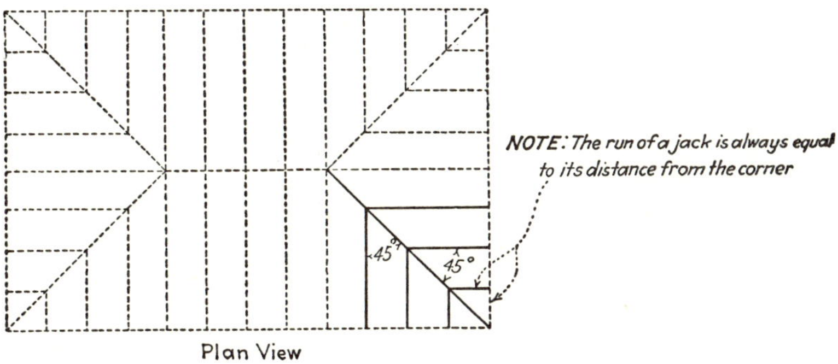

Fig. 44.

As previously shown in the plan view, the hip rafter of an equal pitch is *always* at a 45-deg. angle to the plate line of the building (see Fig. 26, p. 24). As hip jacks are always placed in pairs, the scale run of each pair, and the outside plate lines as drawn in a roof layout, will form a geometrical figure in the shape of a square.

It is obvious that, if the length of *one* side of a square is known, the length of any of the other three sides is the same. Therefore, if the distance that a jack rafter is to be placed from the corner of a building is known, this distance is the same as the run of the opposite jack rafter. From this, the rule is deduced that *the run of a hip jack rafter is always equal to its distance from the corner.*

Length of a Hip Jack by the Line Length Method Based on the Slant Triangle.—(Figure 45.) The step-off methods as just explained are good methods and commonly used in the trade, but they do not give the length of the rafter until *after* the layout process is completed. This sometimes means that the rafter must be laid out a second time as the first tryout may prove that the piece of lumber selected was not quite long enough.

By using the "slant triangle" method, the length of a rafter can be determined before any layout work is done. This slant triangle method was used in finding the line length of the hip rafter and explained on page 25. It can be applied to the jack

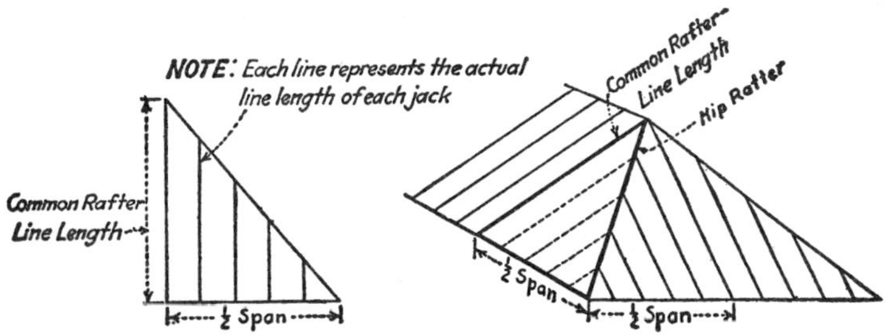

FIG. 45.

by carrying it a step farther. On any flat surface draw a triangle to 1⁄12 scale, one leg of the triangle to represent *one-half the span of the roof* and the other leg at 90 *deg.* to the first to represent the *line length of the common rafter.* The bridge distance between the two ends is the scale line length of the hip rafter. Any line representing a jack rafter drawn in *parallel* to the common rafter will be the scale line length of that jack rafter. This is a very flexible method in that, irrespective of the spacing of the jacks, their length is always equal to the scale length of the line that represents the hip jack.

Length of the Hip Jack by the Line Length Method Based on the Common Difference.[1]—(Figure 46.) The slant triangle method can be used to further advantage when finding lengths of hip jacks, and makes it possible to develop the actual difference between each succeeding jack, as there is a *common difference* in the length of any set of jacks in a particular roof, providing they are *equally spaced,* which is usually the case.

[1] See problem 4, Chap. VI, page 101.

Draw the slant triangle as before but, instead of showing all of the jacks, draw in only the first long one next to the common rafter. The difference between the length of this jack and the common rafter, which can be measured (see x in the drawing), is the common difference between all of the jacks. That is to say, each jack rafter should be shortened the common difference length from the next longer jack, or vice versa.

By developing the common difference it is possible to have several men cut jack rafters at the same time. Each man can be notified what the common difference is and cut the jacks accordingly.

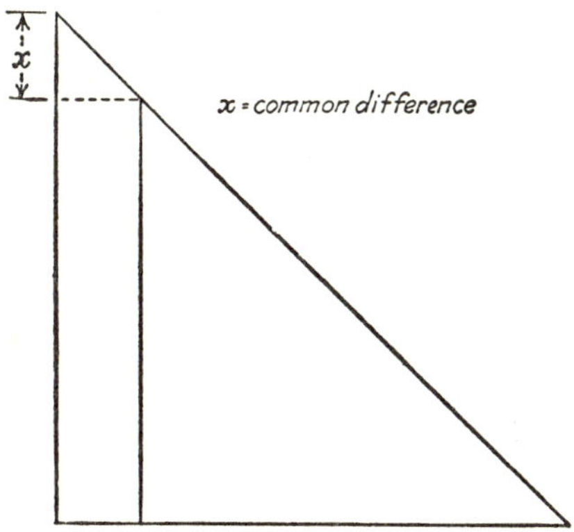

NOTE: Common difference developed by the slant triangle process

FIG. 46.

Length of the Hip Jack Rafter by the Line Length Method Based on the Common Difference Figured Mathematically.—(Figure 47.) The common difference can also be found mathematically so that the slant triangle need not be drawn. If the number of hip jacks that will be placed between the last common rafter and the corner of the building is known, *dividing the line length of the common rafter by one more than the number of jacks* gives the common difference.

As soon as this common difference is determined, measure the distance successively on the top edge of the common rafter pattern, beginning from the top and driving nails at each point located (see Fig. 47). The common rafter pattern will then serve as a pattern for all of the hip jacks and a pair of jack rafters can be cut to each length represented by the different nails. These lengths of course, will be "line lengths" and allowance for shortening will have to be made. This is explained on page 46.

Length of a Hip Jack Rafter by the Line Length Method, Figured by "Diagonal of the Cut."—(Figure 48.) For a quick

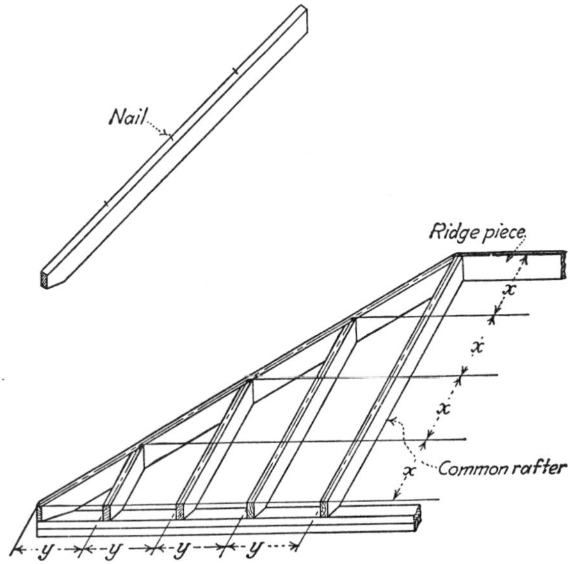

NOTE: *Common rafter line length divided by one more than the number of jacks = common difference*

FIG. 47.

method of finding whether a piece of stock is long enough for a given jack rafter, the following method is probably the best.

With gages set on the square to the "cut of the roof," measure the bridge distance between them. This result is the *actual length of the common rafter for every foot of run.* Jack rafters, being parallel to the common rafters, the length of rafter per foot of run will, of course, be the same.

If the length of a rafter for a foot of run is known, the line length of that rafter will be as many times more than the length for a foot of run as the total run is greater than the unit of run (12 in.).

To illustrate:

For a roof of a 9 and 12 cut, the "diagonal of the cut" is 15 in. A rafter that has 11 ft. of total run will then be 11 × 15 in., which is 165 in., or 13 ft. 9 in. long. For a ⅓-pitch roof of an 8 and 12 cut, the diagonal of the cut is 14⅜ in. For a given total run of 16 ft. the line length will be 16 × 14⅜ in., which is 230 in., or 19 ft. 2 in.

Cut of the Hip Jack.—As the hip jack is parallel to the common rafter, the cut of the jack is always the same as the cut of the common rafter.

In order to clarify the use of the different methods given in the preceding pages, the following outline is presented so as to bring out the fact that, while apparently there are a number of different methods by which the line length of the hip jack rafter can be found, actually there are only *two*. The different processes are only variations of these two.

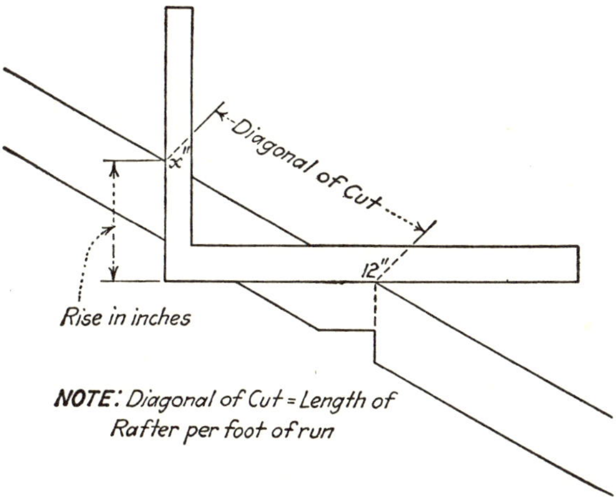

Fig. 48.

The step-off method, based on run....	The graphic method (Fig. 43) The location method (Fig. 44)
The line length method.............	The slant triangle (Fig. 45) The common difference: The slant triangle (Fig. 46) The common rafter (Fig. 47) The diagonal of cut (Fig. 48)

How to Lay Out the Cheek Cut of a Jack Rafter.[1]—(Includes the Shortening, Fig. 49.) The cheek cut of a hip jack rafter is the same shape and laid out in the same way as the cheek cut of a hip rafter. In either case, in the plan view, both the hip and the hip jack rafters are at a 45-deg. angle to the piece to which they are framed. On account of the thickness of the hip rafter, the jack rafters will be shortened one-half of the thickness of the hip rafter. This thickness must be "45-deg. thickness" because, as shown in the *plan view*, the hip jack is at an angle of 45 deg. to the hip (see Fig. 44).

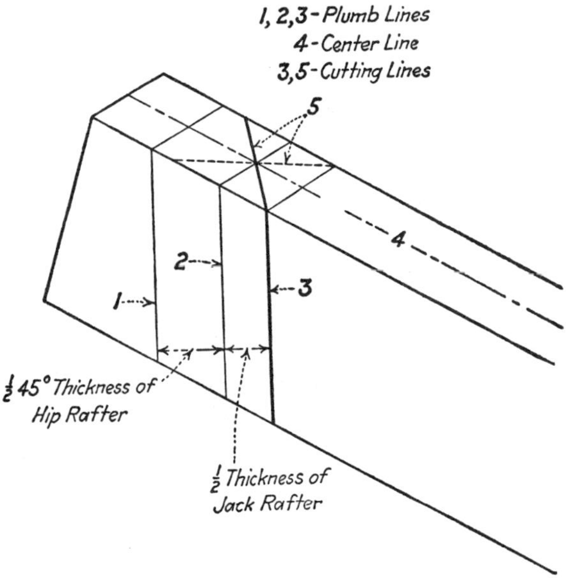

FIG. 49.

The principle of the cheek cut is discussed on page 28 in Chap. III. In laying out any cheek cut, always *measure at right angles to the plumb cut.* For a *full cheek* cut measure the full thickness of the rafter stock. For a *double cheek* cut measure one-half the thickness of the rafter stock. In laying out a double cheek cut the measuring or line length line will always come to the center of the cheek cut line.

[1] See problem 5, Chapter VI, page 105.

If the shortening is done before the cheek cut is laid out, only one cheek cut line will be needed. If, however, the shortening is done after the cheek cut line is marked, two cheek cut lines are necessary. Therefore always shorten first, and then lay out the cheek cut.

The steps in the layout process of a cheek cut of a hip jack rafter, then, are as follows:

1. Draw a plumb line at the top end of the rafter stock. The cut of the jack will be the same as the cut of the common rafter.

2. To allow for the shortening, measure *at right angles* to this plumb line a distance equal to one-half the 45-deg. thickness of the hip rafter stock. Draw another plumb line at this point.

3. Measure at *right angles* to this second plumb line a distance equal to one-half the thickness of the jack rafter stock and draw a third plumb line.

4. Draw a center line down the top edge of the jack rafter.

5. Draw angle lines as shown in Fig. 49 connecting the last plumb line drawn with the intersection of the center line and the second plumb line when squared across the top edge of the rafter. The cutting is done on lines (3) and (5).

The direction of the angle lines is determined by the position of the rafter in the roof. Since hip jacks are always cut in pairs, the angle line for an average roof can be drawn in either direction, as rafters are required with the cut in both directions.

Seat Cut, Overhang, and Tail Cut of a Hip Jack Rafter.— As the seat cut, overhang, and tail cut of the hip jack are exactly the same as the corresponding parts of the common rafter, use the common rafter pattern in laying out the jack. The line length line of the jack, as measured from the cheek cut, will locate the position of the seat cut.

HIP ROOF RIDGE LENGTH (FIGURE 50)

There are two lengths on a hip roof ridge: first, the *theoretical length,* which is always the difference between the width and length of the building; and, second, the *actual length* of the ridge. The actual length is based on the type of construction used as shown in Fig. 23 (p. 22), and is always *longer* than the theoretical length.

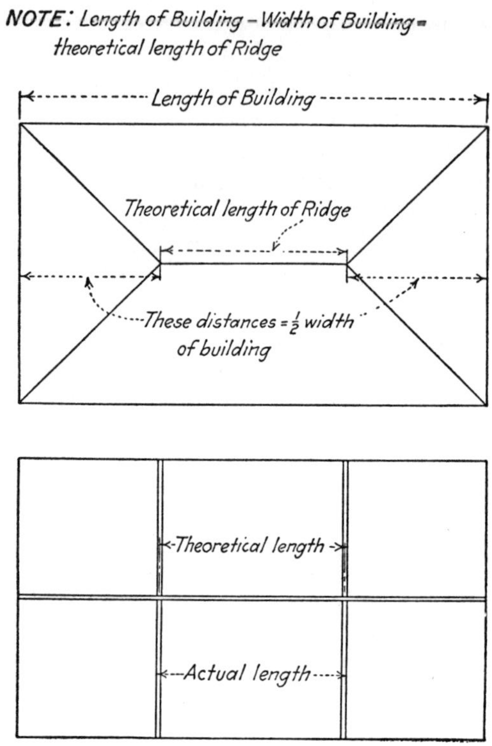
NOTE: Length of Building – Width of Building = theoretical length of Ridge
Length of Building
Theoretical length of Ridge
These distances = $\frac{1}{2}$ width of building
Theoretical length
Actual length

FIG. 50.

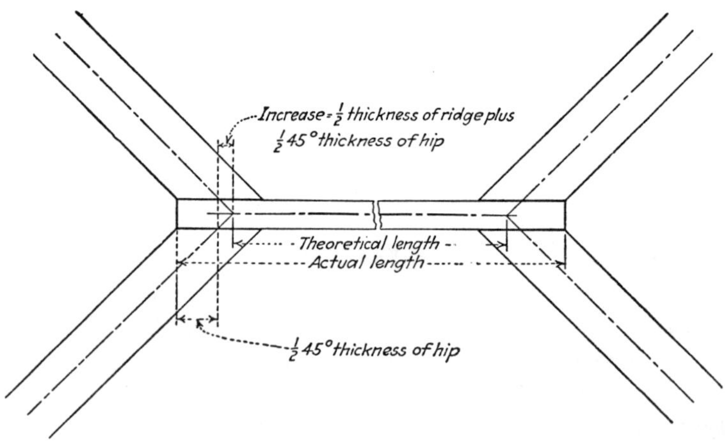
Increase = $\frac{1}{2}$ thickness of ridge plus $\frac{1}{2}$ 45° thickness of hip
Theoretical length
Actual length
$\frac{1}{2}$ 45° thickness of hip

FIG. 51.

Figure 51 illustrates the actual length of the ridge when the hip rafter is framed to it. For this type of construction the

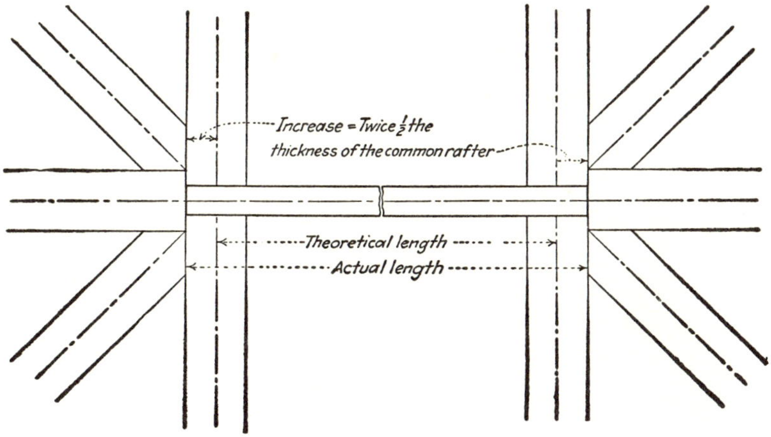

Fig. 52.

theoretical length of the ridge must be increased the thickness of the ridge stock and the 45-*deg*. thickness of the hip rafter.

If the hip rafter is framed to the common rafters, as shown in Fig. 52, add to the theoretical length of the ridge the thickness of the common rafter stock.

Note: If the roof is hipped at one end only, the ridge length is increased *one-half* the amounts described above.

CHAPTER IV

THE INTERSECTING ROOF

I. The Supporting Valley:

- Cuts and terms
- Unit of run
- Line length
- Ridge cut
- Seat cut
- Overhang
- Tail cut

II. The Shortened Valley:

- Cuts and terms
- Line length
- Ridge cut

III. The Jack Rafter:

- *a*. The regular valley jack:
    - Line length
    - Ridge cut
    - Cheek cut
- *b*. The cripple jack:
    - Line length
    - Cheek cut
    - Layout process
- *c*. The hip-valley jack:
    - Line length
    - Cheek cuts
    - Layout process

IV. The Ridge:

- Length of ridge when ridge boards are the same height
- Length of ridge board when ridges are of a different height

CHAPTER IV

THE EQUAL PITCH INTERSECTING ROOF OF A SQUARE-CORNERED BUILDING

An intersecting roof is one which is a combination of both a gable and a hip roof and is so-called on account of the shape of the building on which it is to be framed. Any building that is rectangular in shape will have either a flat, gable, or hip roof. A building, however, that has a wing or wings added to it requires an additional roof for each section. Where any two roofs meet or intersect, several new types of rafters are required.

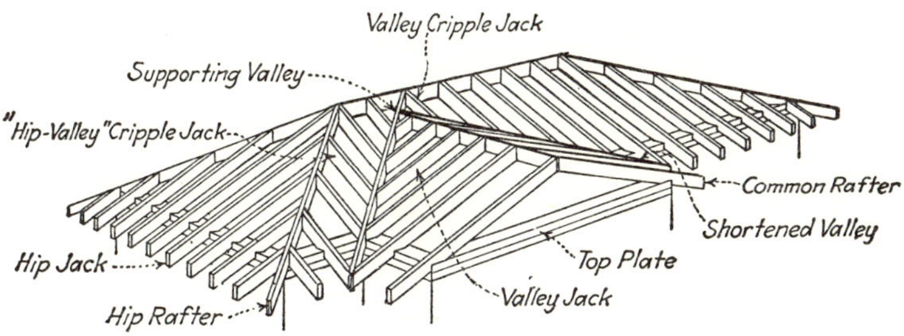

FIG. 53.

These new rafters are of two classes: valley rafters and valley jack rafters. The valley jack rafters are again classified into three types or kinds: the valley jack, the cripple jack, and the hip-valley jack (see Fig. 53).

No mention need be made of the common rafters, hip rafters, and hip jack rafters as these have been fully discussed in the previous chapters.

THE VALLEY RAFTER

The valley rafter is the rafter that lies at the intersection of the planes of the two sections of the roof. On account of the difference possible between the width of the two spans of a building there are really *two* kinds of valleys. The first, or *supporting* valley, is framed from the plate to the main ridge and supports the second, or *shortened valley* which "dies" against it.

Cuts and Terms of a Supporting Valley.—The cuts of a valley rafter are similar in name to those of any other rafter: namely, the *ridge, seat,* and *tail* cuts. The part of the rafter that extends beyond the building is likewise called the *overhang*.

Figure 54 illustrates the order of operations in framing a valley rafter. The line length is found first and then, working

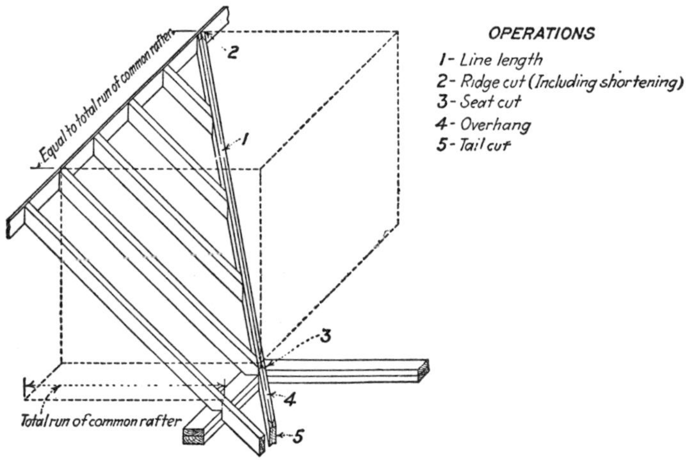

FIG. 54.

down from the top end in succession, the ridge cut, the seat cut, the overhang, and the tail cut are laid out.

Unit of Run.—(Figure 55.) The unit of run for the valley rafter is the same as the unit of run for the hip rafter, as the valley rafter in *plan view* is always at an angle of 45 deg. to the plate line of the building (Fig. 26). As 17 in. is the diagonal of a square 12 × 12 in., 17 in. is the unit of run.

Cut of the Valley Rafter.—(Figure 56.) The cut of the valley rafter is identical with the cut of the hip rafter, that is, "rise in inches and the valley unit of run (17 in.)."

Line Length of the Valley Rafter.—(Figure 57.) The slant triangle method as developed in the chapter on the hip rafter

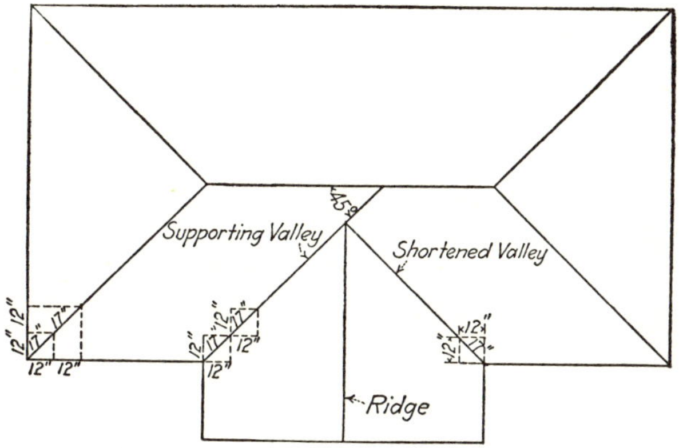

FIG. 55.

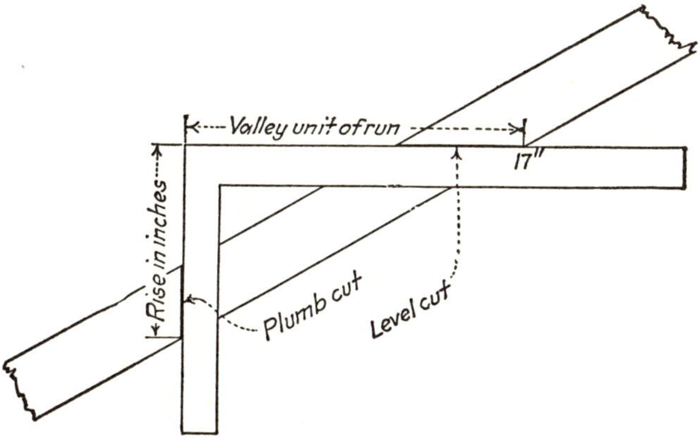

FIG. 56.

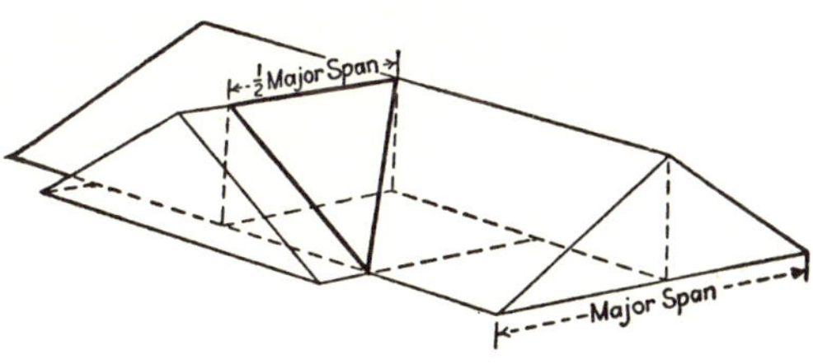

FIG. 57.

can also be applied to the valley rafter. The valley rafter is the hypotenuse of a triangle formed by *one-half the span and the line length of the common rafter.* If the gages are set on the steel square to the scale of these dimensions, measuring the bridge distance gives the scale line length of the valley rafter.

The Ridge Cut of the Supporting Valley Rafter.—(Figures 58 and 59.) The ridge cut of the supporting valley rafter is a *single cheek* cut when framed to the ridge (see Fig. 58). If the

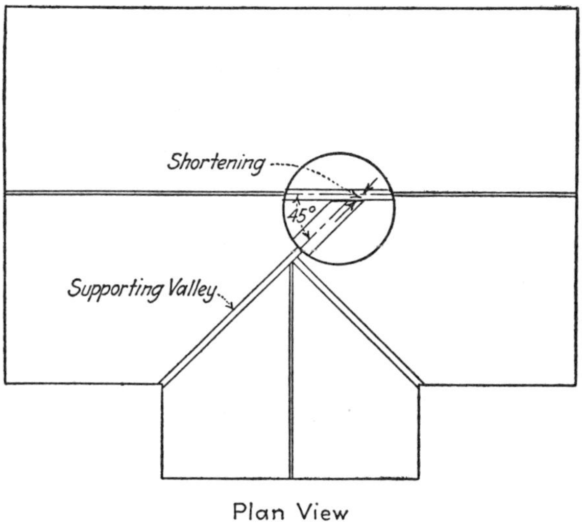

FIG. 58.

ridge heights of both parts of the building are the same, the supporting valley ridge cut will be a *double cheek* cut (see Fig. 59). The layout process of both single and double cheek cuts is the same. The shortening of the valley rafter to allow for the ridge stock is one-half the 45-deg. thickness of the ridge material.

Layout Processes.—(Figure 59.) The drawing illustrates the layout of the ridge cut. The order of procedure is as follows:

1. Draw a plumb line near the top end of the rafter stock.

2. Measure at *right angles* to this line the proper allowance for shortening and draw a second plumb line.

3. Measure at *right angles* to this second plumb line a distance equal to one-half of the thickness of the valley rafter stock.

4. "Square" all three lines across the top edge of the rafter and draw in the angle lines of the cheek cut.

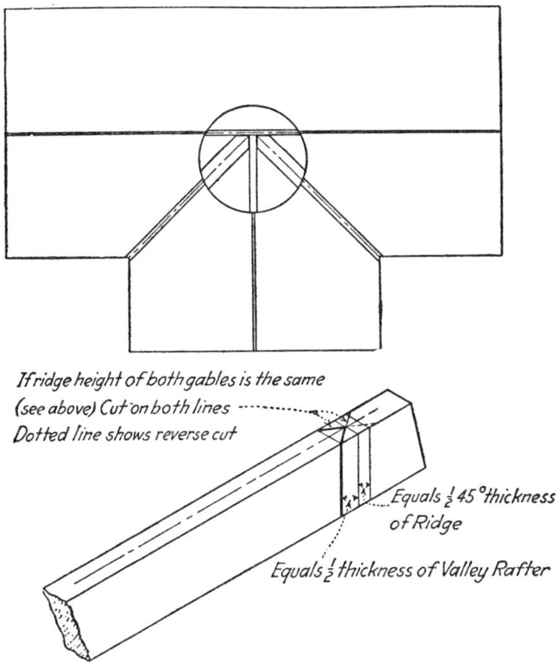

Fig. 59.

The Seat Cut of the Valley Rafter.—(Figure 60.) The seat cut of the valley rafter is very much like the seat cut of the hip rafter, although *reverse* in shape. The *V*-shaped cut is *below* the line length line of the rafter. No "drop" need be allowed for as the sheathing "strikes" the valley rafter on its center line. To lay out a valley rafter seat cut, proceed in the following manner:

1. Draw a plumb line at the line length distance from the ridge cut.

2. Measure at *right angles* to this line and *towards* the bottom end of the rafter a distance equal to one-half the thickness of the valley rafter stock and draw a second plumb line.

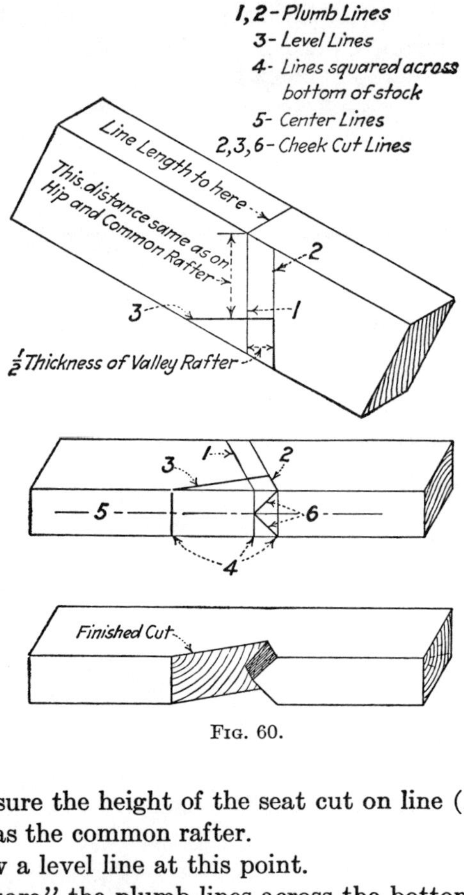

FIG. 60.

3. Measure the height of the seat cut on line (1), making it the same as the common rafter.

4. Draw a level line at this point.

5. "Square" the plumb lines across the bottom edge of the rafter.

6. Connect the center of line (1) with the edge of line (2). These angle lines are the cutting lines of the seat cut together with the level line. The resultant cut will exactly fit into the angle of the building.

The Overhang of the Valley Rafter.—(Figure 61.) The overhang length of the valley rafter is based on its run as in the case of the hip rafter (see p. 36). The *run* of the valley is always the diagonal of the run of the cornice for the common rafter. Therefore:

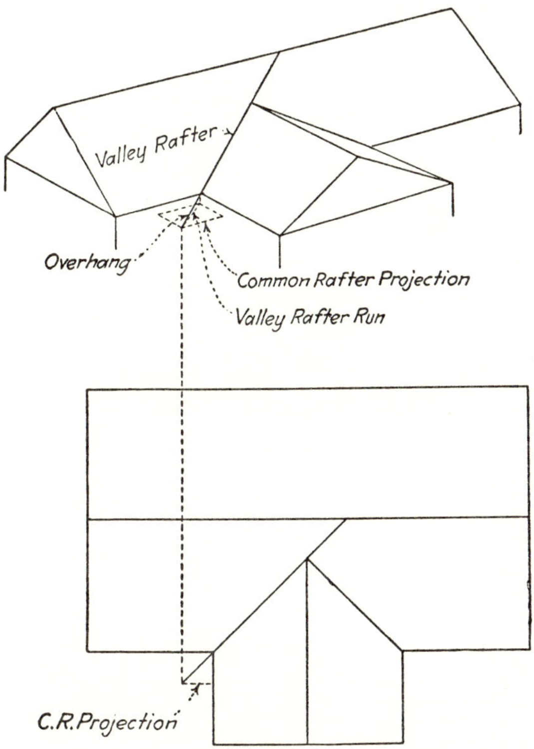

Fig. 61.

1. Find the diagonal (45-deg. angle) of the cornice run of the common rafter. This will equal the valley rafter run.
2. When run of valley is known, step off as many times as necessary to lay off that run, using the cut of the valley (rise in inches and 17).

The Tail Cut of the Valley Rafter.—(Figure 62.) It should be very apparent by this time that the valley rafter cuts are laid off exactly as are the hip rafter cuts *with the exception* that the direction of measurement is *exactly opposite.* Therefore, for the valley rafter tail cut:

1. Measure at *right angles* to the plumb line that marks the extreme line length of the rafter and *toward* the lower end a distance equal to *one-half its own thickness.*

2. Square both plumb lines across the top edge of the rafter stock.

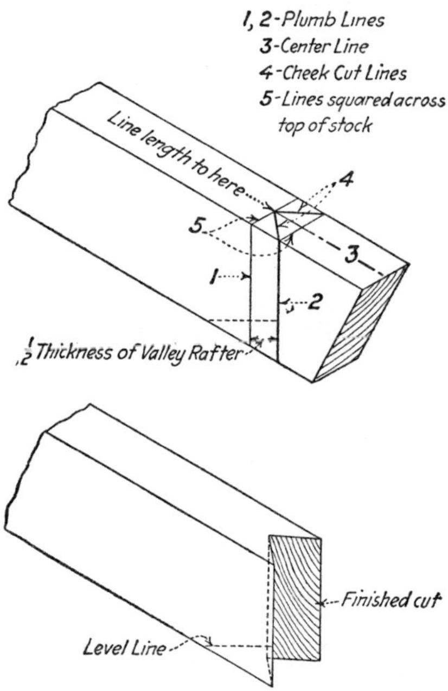

FIG. 62.

3. Draw diagonal lines as shown. The shape of this angle permits a piece of fascia to fit into it from both parts of the building.

4. If desired, draw a level line, using the square and marking on the 17-in. side of the blade. Cut on angle lines and level line.

Cuts and Terms of the Shortened Valley.—(Figure 63.) The shortened valley rafter is so called because it is shorter than the supporting valley rafter to which it is framed. It is shorter due to the fact that while the supporting valley rafter is continued from the plate line to the ridge of the main roof, the shortened valley rafter continues from the plate line up to the supporting valley only. This point of intersection determines the ridge height of the minor span.

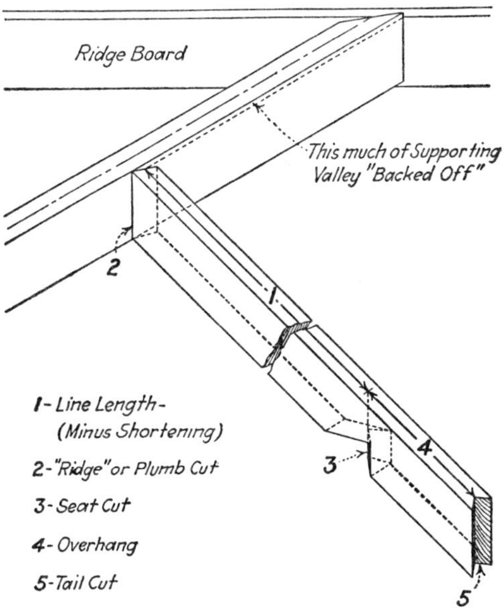

Fig. 63.

The cuts and terms of the shortened valley are identical with those of the supporting valley with the one exception of the ridge cut. The layout order of procedure is the same:

1. Determine the line length.
2. Lay out the ridge cut (allowing for shortening).
3. Lay out the seat cut.
4. Determine its run and lay out the overhang.
5. Lay out the tail cut.

Note that the shortened valley ridge cut is not the same shape as the supporting valley ridge cut.

Line Length of the Shortened Valley.—(Figure 64.) The drawing illustrates the use of the slant triangle in finding the line length of the shortened valley rafter.

Rule: Find the diagonal of the line length of the minor roof common rafter and one-half the *minor span.*

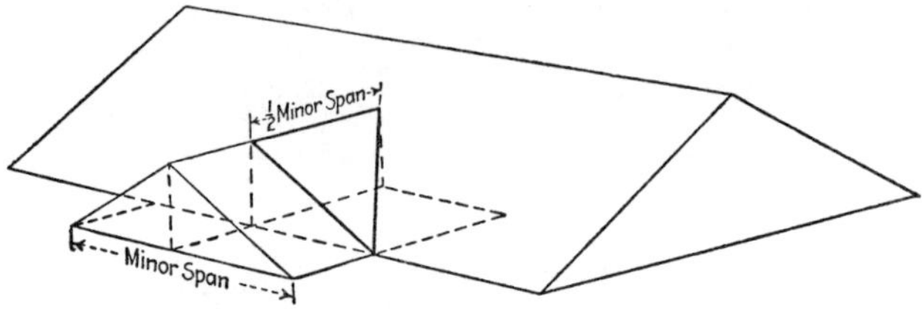

FIG. 64.

The Ridge Cut of the Shortened Valley Rafter.—(Figure 65.) The ridge cut of the shortened valley rafter is exactly the same *shape* as the ridge cut of a common rafter. The plan view

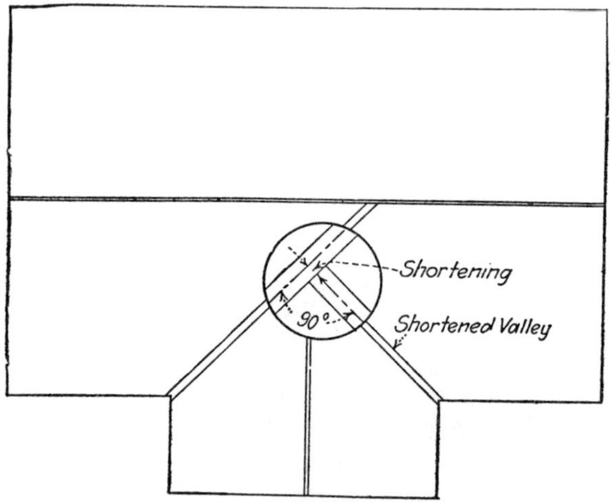

Plan View
FIG. 65.

shows the two valleys to be at an angle of 90 deg. to each other. The shortening of this rafter is also one-half the square thickness of the supporting valley rafter.

How to Lay Out the Ridge Cut.—(Figure 66.) 1. Draw a plumb line near the top end of the rafter stock, using "rise in inches" and 17 as the cut.

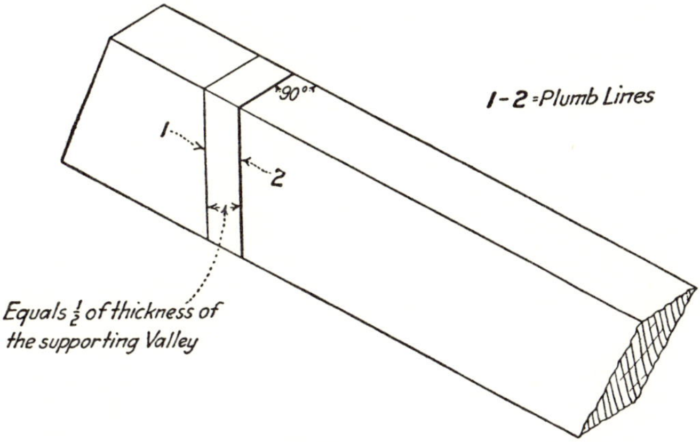

Fig. 66.

2. Measure at right angles to this line and toward the bottom end of the rafter a distance equal to one-half the thickness of the supporting valley rafter stock.

3. Square this second line across the top edge of the stock and cut the rafter on this line.

Note: Intersecting roofs can be constructed without using valley rafters. The main roof is sheathed as though the building had no wing. The valley is formed by nailing a 1 × 4 in. (or wider) on *top* of the sheathing of the main roof, the top edge of which is made to "line up" with the top edge of the rafters of the wing. Jack rafters are cut to fit on top of the sheathing, requiring a level cheek cut as shown in Fig. 106 (page 107). The smaller roof is then sheathed as usual. The cutting angle of the sheathing to make it fit to the valley is "run and length, cut on run."

This type of construction, known as a "*blind valley*," is very sturdy, as the minor roof is actually framed on top of the main roof.

THE VALLEY JACKS

Figure 67 illustrates the three kinds of valley jack rafters. The cheek cuts of each rafter as shown are laid out in the usual way (see p. 30). Their line lengths, however, can be determined by a number of different methods, each of which is discussed separately.

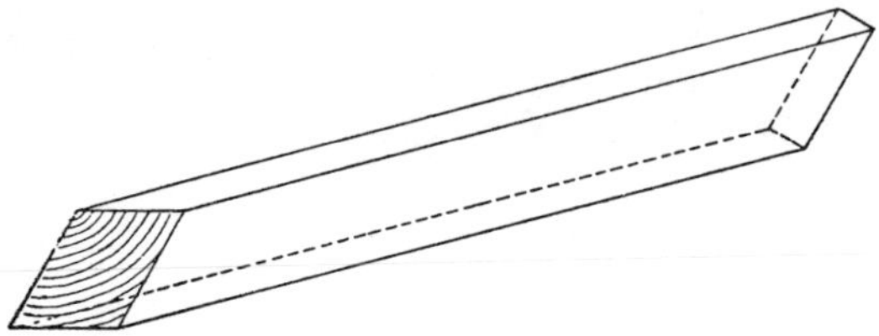

Valley Jack

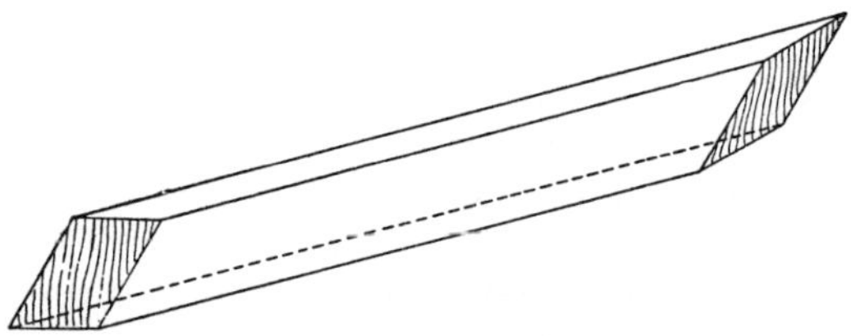

Cripple Jack

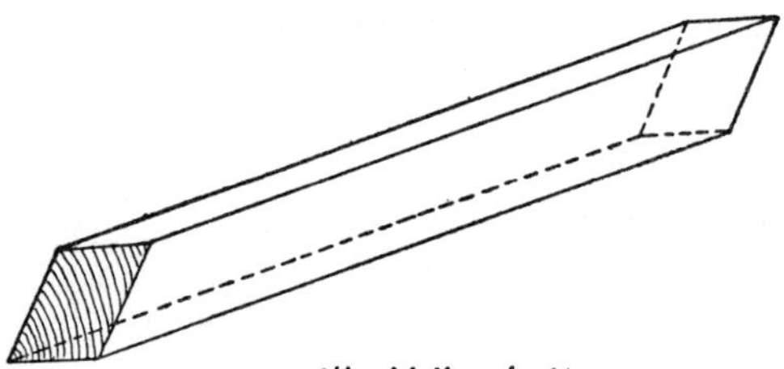

Hip-Valley Jack

FIG. 67.

The Valley Jack Rafter.—(Figure 68.) The valley jack is laid out in the usual way and in the same order as for any rafter.

1. Find the line length.
2. Lay out the ridge cut.
3. Lay out the cheek cut.

How to Find the Line Length of the Valley Jack Rafter.—(Based on Run, Fig. 69.) The positions of the various

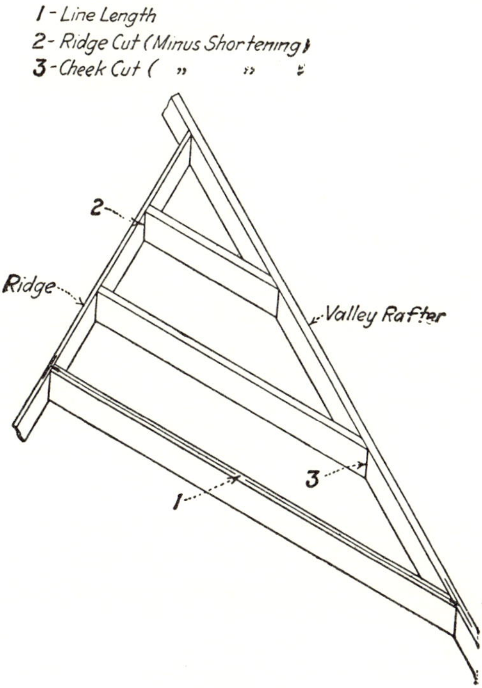
1 - Line Length
2 - Ridge Cut (Minus Shortening)
3 - Cheek Cut (" ")
2
Ridge
Valley Rafter
3
1

FIG. 68.

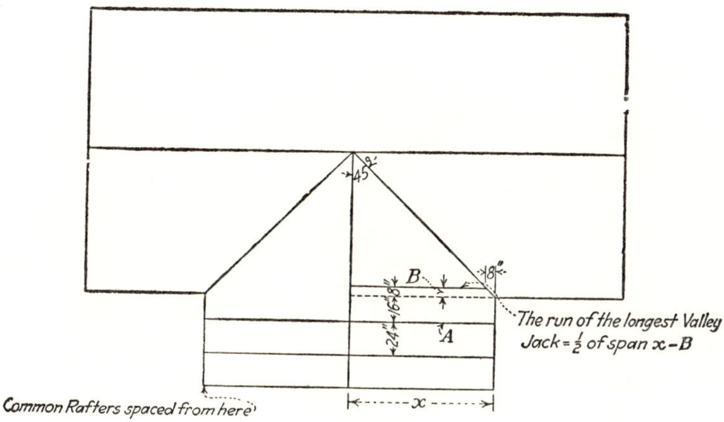
45°
B
8"
8"
16"
24"
A
The run of the longest Valley Jack = ½ of span x - B
x
Common Rafters spaced from here

FIG. 69.

rafters in a roof are, as a general rule, spaced beginning from the corner of a building. This is especially true of the valley jacks. The *run* of the longest valley jack can be determined if the distance that it is to be located from the angle of the building is known. The run of the longest jack is always "one-half the minor span minus the difference between the center of the last common rafter and the inside angle of the building and the regular rafter spacing."

To illustrate:

Assume a run of 11 ft. and a rafter spacing of 24 in. The last common rafter is placed 16 in. from the inside angle of the building. The run of the longest jack is 11 ft. minus the difference between 24 in. and 16 in., or 11 ft. minus 8 in., which is 10 ft. 4 in. When the run is known, use the step-off method and lay out the jack. The plumb and level cuts are the same as those of the common rafters to which they are parallel.

Line Length of the Valley Jack by Location.—(Figure 70.) The distance that any jack rafter may be placed on the ridge

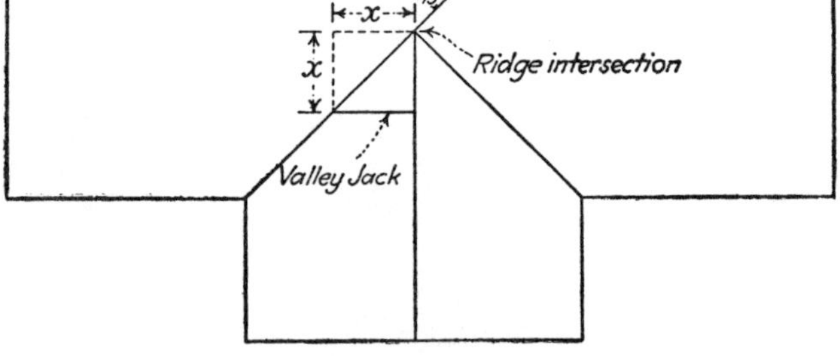

FIG. 70.

piece from the ridge and valley intersection *is equal to the run of* that rafter. Therefore, use the step-off method for finding its line length.

Line Length of the Valley Jack by Common Difference.—(Figure 71.) Draw to scale two legs of a slant triangle at 90 deg. to each other, one to represent one-half the span, and the other, the line length of the common rafter. Complete the triangle. The line last drawn represents the line length of the valley rafter.

Next, represent the first long valley jack rafter by a line drawn parallel to the common rafter line and find the difference X as shown in the drawing. Since X equals the common difference, each succeeding valley jack is cut this much shorter. For a fuller explanation of the common difference methods, see page 42.

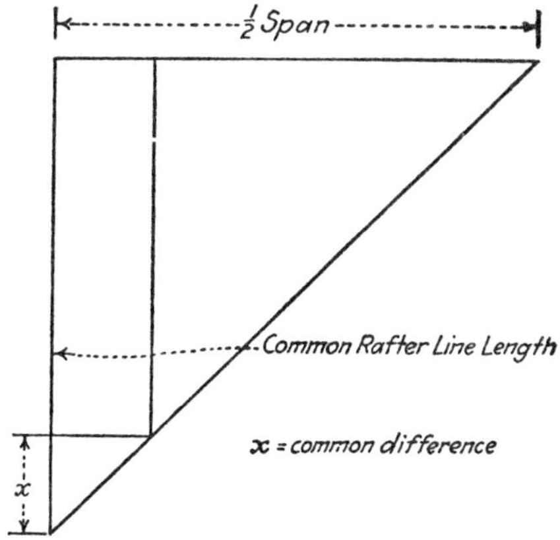

FIG. 71.

Ridge and Cheek Cuts of the Valley Jack.—(Includes Shortening, Fig. 72.) The completed layout of the valley jack rafter is illustrated in the drawing. It will be noted that the rafter has been shortened at both ends, at the ridge end for the ridge stock and at the cheek cut end for the valley rafter. In actual practice the amounts to shorten can be added together and deducted from one end. As the layout process is

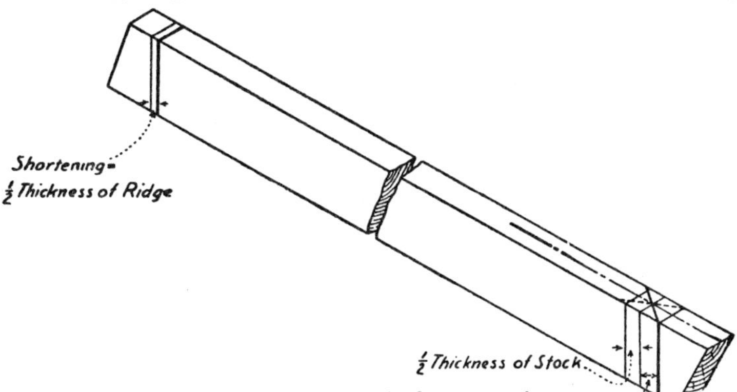

FIG. 72.

somewhat simpler at the ridge end of the rafter, it is advisable to do the shortening there.

The Valley Cripple Jack Rafter.—(Figure 73.) The valley cripple jack rafter is the rafter that is framed between the shortened valley and the supporting valley. Each end has a

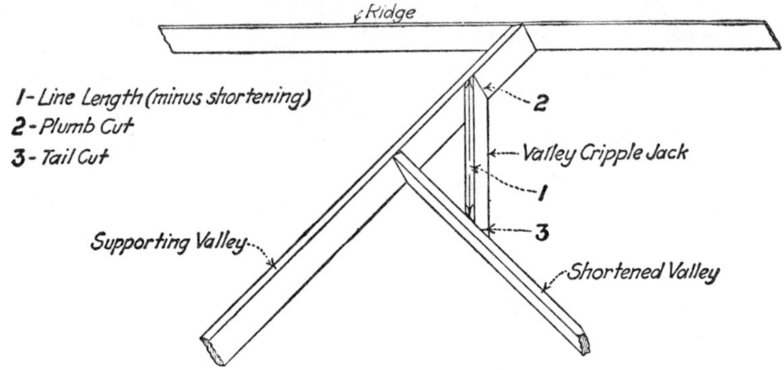

FIG. 73.

cheek cut, the angle of the cuts being in *opposite* directions. The layout procedure is as follows:

1. Determine the line length.
2. Lay out the top end cut.
3. Lay out the bottom end cut.

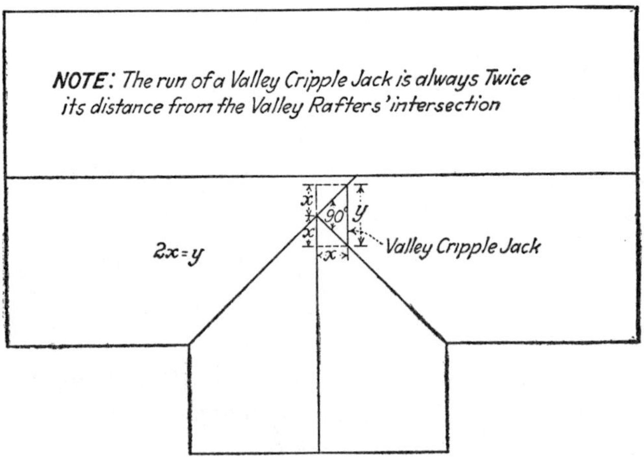

FIG. 74.

The Line Length of the Valley Cripple Jack Based on Run.—(Figure 74.) Finding the run of the valley cripple jack is a comparatively easy matter if the distance of the rafter from the valley intersection is known. It will be noted that the "run of the valley cripple jack is always *twice* its distance from the valley rafter intersections."

To illustrate:

If the distance X is 3 ft., the run of the cripple jack is twice 3 ft., or 6 ft. This is always true because of the 90-deg. angle formed by the position of the two valley rafters, as shown in the plan view.

Cheek Cuts of the Valley Cripple Rafter.—(Includes Shortening, Fig. 75.) The cheek cuts of the valley cripple jack

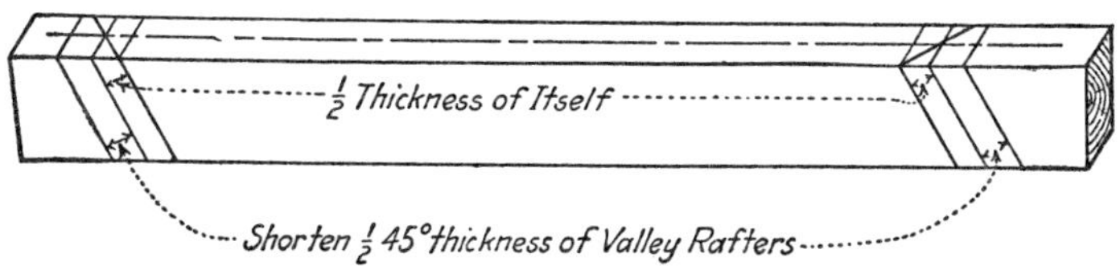

Fig. 75.

are regular cheek cuts and are laid out in the regular manner. There are *two* shortenings to allow for. If the two valley rafters are the same thickness (and this is usually the case), instead of allowing for the shortening twice deduct the 45-deg. thickness of the valley rafter in one operation

How to Lay Out the Valley Cripple Jack.

1. Find the run of the cripple jack.
2. Use step-off method and lay out, using the cut of the common rafter for the angle of the cuts.
3. Lay off the regular cheek cuts making proper allowance for the shortening. Be sure that the cheek cuts are in *opposite* directions.

The Hip-valley Cripple Jack.—(Figure 76.) The hip-valley cripple jack is the rafter that is framed between a hip and valley rafter and is always required when distance A (in the figure) is *less* than one half of the span of the building As the hip and valley rafters are always *parallel* to each other in the equal pitch roof, it follows that every jack that is framed between these two rafters will always be the same length.

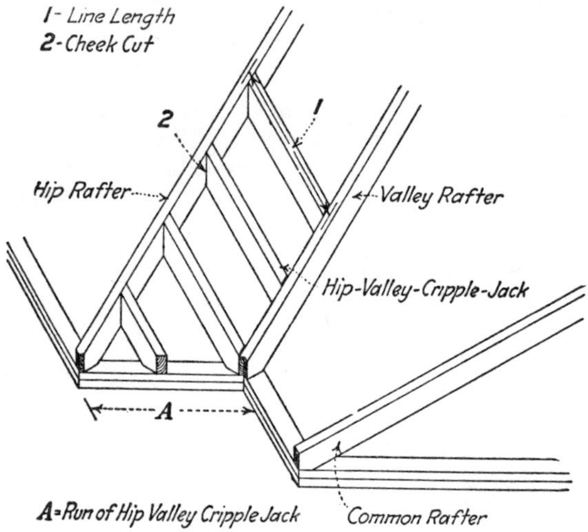

FIG. 76.

Line Length of the Hip-valley Jack Rafter Based on Location.—(Figure 77.) The run of the hip-valley cripple rafter

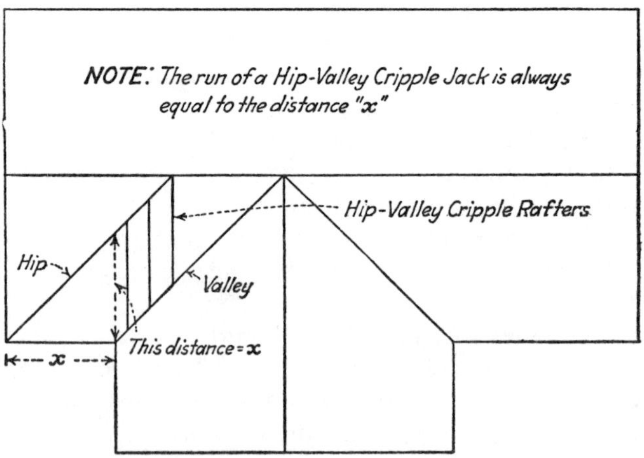

FIG. 77.

is always equal to the distance between the seat cuts of the hip and valley rafters to which the jacks are framed (see x in the drawing). This rule corresponds to the rule used in finding the run of the hip jack as described on page 40, Chap. III.

Cheek Cuts of the Hip-valley Cripple Jack.—(Figure 78.) The cheek cuts of this rafter are the usual cheek cuts. Note that the cutting lines are *parallel* to each other.

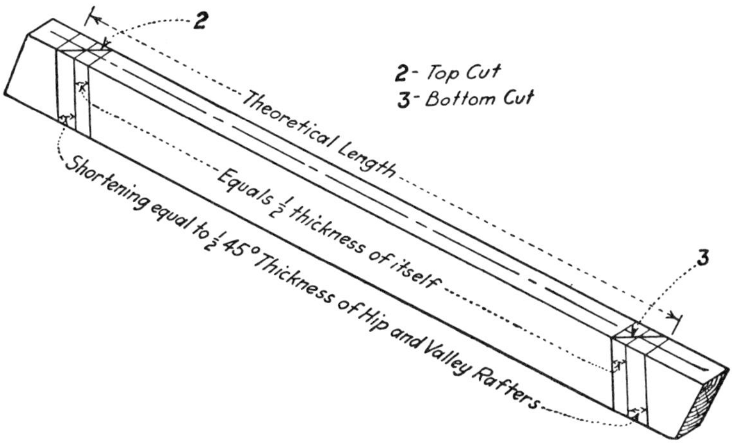

FIG. 78.

How to Lay Out the Hip-valley Cripple Jack Rafter.

1. Find the run of the rafter.

2. Use step-off method and lay out, using the cut of the common rafter as the cut of the jack rafter.

3. Lay out the cheek cuts, making allowances for shortening at each end of the rafter. (This is of course one-half the 45-deg. thickness of both the hip and valley rafters.) This shortening can be combined and deducted from one end.

THE RIDGE OF THE INTERSECTING ROOF

There are two problems connected with the lengths of the ridge of an intersecting roof, one where the ridge heights are equal and the other where the intersecting roof ridge drops below that of the main roof.

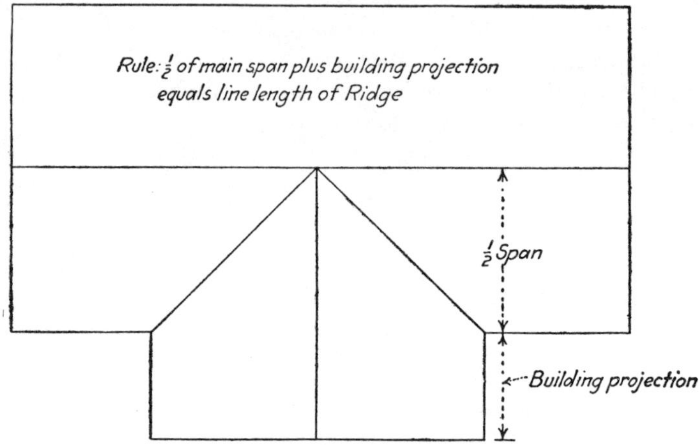

FIG. 79.

Ridge Length When Ridges Are the Same Height.—(Figure 79.) The theoretical ridge length (when the ridge heights

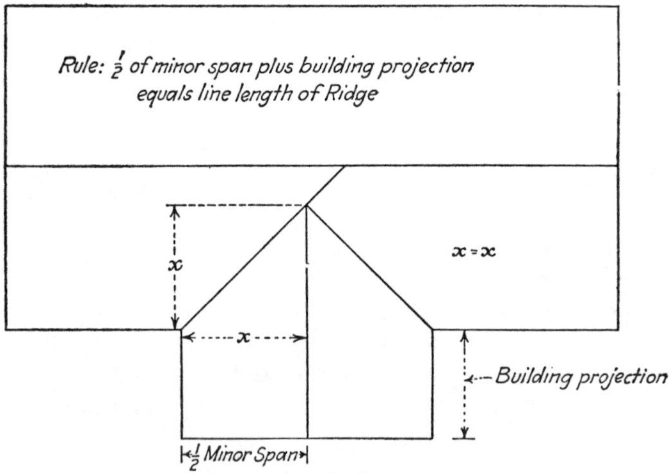

FIG. 80.

are the same) is equal to the building *projection plus one-half the main span.* The actual length is shorter due to the fact that the one ridge is framed to the other. In common practice it is customary to make the intersecting ridge a little *longer* than its theoretical length, which permits the plumbing of the common rafters at the gable line. As any surplus stock can be sawed off after the roof is raised, it is advisable to allow for more than the theoretical length.

The shape of the ridge at the intersection is simply a *square plumb cut.*

Ridge Length When the Ridges Are Not the Same Height.—(Figure 80.) The theoretical length of the ridge when the ridges are not the same height is equal to the *building projection plus one-half the minor span.* As before suggested, the ridge should be cut longer to allow for the plumbing of the outside pair of common rafters.

The shape of the end of the ridge at the point of intersection would really be a double 45-deg. plumb cut. In common practice, however, and for all practical purposes, the stock is usually cut square, as it is not necessary to make the ridge piece fit exactly into the valley rafter angles.

CHAPTER V

THE UNEQUAL PITCH INTERSECTING ROOF WITHOUT A CORNICE

I. The Common Rafter:

- Cut
- Line length
- Layout process

II. The Valley Rafter:

- Line length
- Cut
- Layout process
- Ridge cut
- Seat cut
- Shortening

III. The Valley Jack:

- Line length
- Cut
- Layout process
- Shortening
- Cheek cut

THE UNEQUAL PITCH ROOF WITH A CORNICE

I. Increased Total Rise
Increased Total Run
How much to Raise the Top Plate

II. The Valley Rafter:

- Cut
- Line length
- Tail cut
- Seat cut

CHAPTER V

THE UNEQUAL PITCH INTERSECTING ROOF

An unequal pitch roof (Fig. 81) is an intersecting roof in that two spans are involved. The underlying principles of framing, however are not the same because each span has a different pitch.

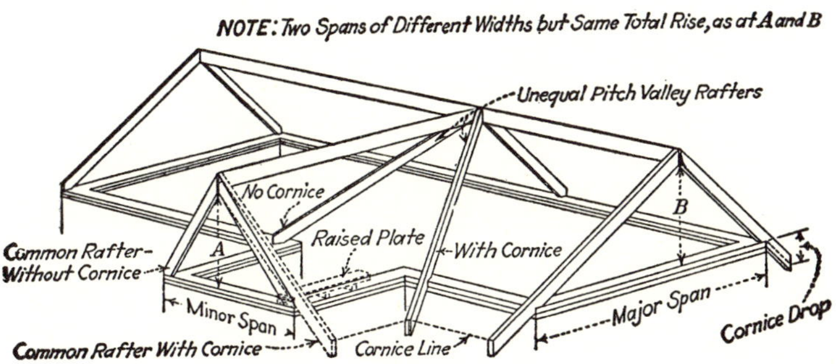

FIG. 81.

The term "*unequal*" is derived from the fact that the distance that each roof "rises" per foot of run is *not* the same. To illustrate:

Assume that the wide or *major* span is 32 ft. and that the pitch for that part of the roof is ⅓. The total rise is one-third of 32 ft., or 10 ft. 8 in. The *cut* for that part of the roof would be the cut for a ⅓-pitch roof, that is 8 in. and 12 in. Note that in the drawing the total rise for the narrow or *minor* span is the same as for the major span, that is 10 ft. and 8 in. Assuming that the minor span is 18 ft., it is easy to see that the cut would not be 8 and 12 but a different one, based on 10-ft. and 8-in. rise and 9-ft. run. The cuts are unequal, hence the term.

It is of course not necessary that the total rise of both spans be the same. In general practice, however, this is usually the case, as one of the reasons for the unequal pitch is the designer's wish to keep the ridge of the building the same height throughout, irrespective of the width of the spans.

Type of Rafters in the Unequal Pitch Roof.—No new rafters appear in this type of roof, but some entirely new layout processes must be developed in order to master the problems involved in framing the unequal pitch roof.

The problem divides itself into *two* parts: an unequal pitch roof *without* a cornice, and an unequal pitch roof *with* a cornice. Each is discussed separately and in detail.

THE UNEQUAL PITCH ROOF WITHOUT A CORNICE

The Common Rafter.—The common rafter of the unequal pitch roof is exactly the same shape after the cuts are made as any common rafter (Fig. 82). The difference between the

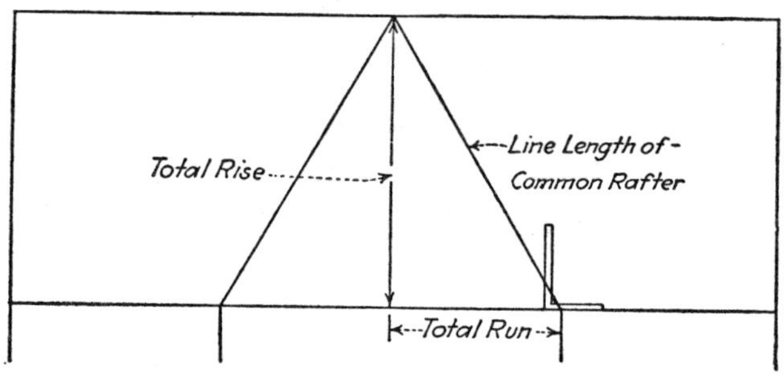

FIG. 82.

equal pitch and the unequal-pitch common rafter is in the method of determining the angle of the plumb and level lines.

Cut of the Common Rafter.—(Without Cornice, Fig. 82.) The problem of finding the cut of the common rafter of an unequal pitch roof is comparatively simple if the *total rise* and the *total run* of the unequal pitch common rafter are used to find the correct angle of the layout lines.

To continue the illustration on page 73, the *cut* of the common rafter for the minor span (using $\frac{1}{12}$ scale) would be

$10\frac{8}{12}$ in. on the tongue of the square to represent the total rise, and 9 in. on the blade to represent the total run.

To Find Line Length of the Common Rafter.—With the gages set to the figures given in the preceding paragraph step off *twelve* times to find the line length of the rafter. It is of special importance to note that when total rise and total run are used, the square is always shifted *twelve* times to find line length, irrespective of the amount of run involved. It is obvious that if, as in this case, 9 in. is one-twelfth of 9 ft., only 12 times 9 in. could equal 9 ft.; therefore step off 12 times. Measuring the bridge distance from gage to gage will also give the scale line length of the common rafter.

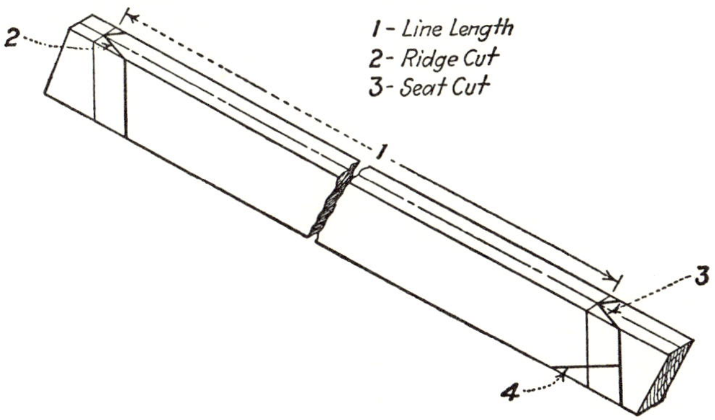

FIG. 83.

Layout for the Cuts of the Common Rafter.—No detailed description is necessary for the ridge and seat cuts of the common rafter for the unequal pitch roof since, aside from the figures used on the square to determine the correct shape of the cuts, the layout processes are identical with those for the ordinary common rafter. This was discussed in detail in Sec. 1 of Chap. II.

The Unequal Pitch Valley Rafter.—The layout process of the unequal pitch valley rafter (Fig. 83) is the same as that of any other rafter:

1. Find the line length.
2. Lay out the ridge cut.
3. Lay out the seat cut.

Line Length of the Unequal Pitch Valley Rafter.—(Based on Total Rise and Total Run, Fig. 84.) The unequal pitch valley rafter is laid out in a manner similar to that of the common rafter described in Sec. 1 of this chapter. Note, however, that the unequal pitch valley rafter and its corresponding common rafter unit of run *never* have the same fixed relation

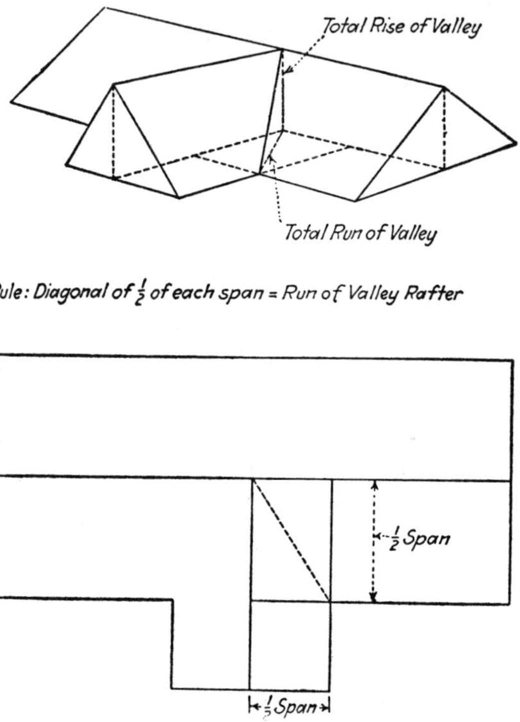

FIG. 84.

(17 to 12) true of the similar rafters in a regular pitch intersecting roof.

The *total rise* of the unequal pitch valley rafter in this example is the same as for the balance of the roof. Using the illustration previously given on page 73, the total rise of the valley *without the cornice* is 10 ft. and 8 in. The total run of the valley is the diagonal of one-half of each span as shown in the drawing.

How to Find the Line Length of an Unequal Pitch Valley Rafter.—To find the line length of an unequal pitch valley rafter proceed as follows:

1. Find the total rise of the roof.
2. Measure the diagonal or "bridge distance" of one-half the two spans involved to determine the run of the valley.
3. Set the gages on the square to the total rise and the total run.
4. Measure the distance from gage to gage to determine the scale line length of the valley rafter.

Cut of the Valley.—As the cut of the valley rafter is "total rise and total run," if the gages are set on the square to these values when line length is being determined, the square will also be ready for use in laying out the correct angles of the cuts. If desired, instead of finding the line length by measuring the bridge distance, use the step-off method, stepping off twelve times.

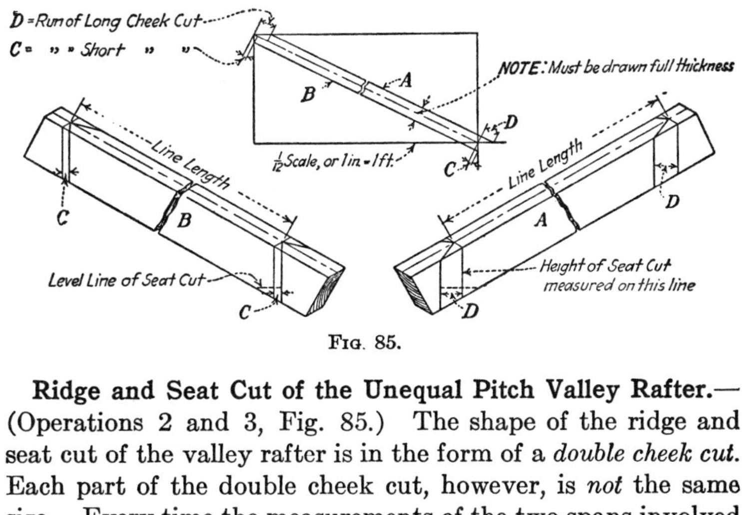

FIG. 85.

Ridge and Seat Cut of the Unequal Pitch Valley Rafter.—(Operations 2 and 3, Fig. 85.) The shape of the ridge and seat cut of the valley rafter is in the form of a *double cheek cut.* Each part of the double cheek cut, however, is *not* the same size. Every time the measurements of the two spans involved are changed, the shape of each part of the double cheek cut also changes.

New Principle Involved in Finding Angle of Cheek Cuts.—(Figure 85.) The new principle involved in finding the unequal pitch rafter cheek cut requires some explanation. The fundamental difference between the runs of the equal and the unequal pitch valley rafter is that the equal pitch run is *always* the diagonal of a *square*, that is, 45 deg., while the unequal pitch is the diagonal of a *rectangle* and never 45 deg. Because this is true, the quickest layout method for the unequal pitch roof is to draw a plan view of the rectangle over which the rafter passes. This must always be drawn to suit the roof to be framed. This plan view is a rectangle drawn to the $1/12$ scale, one side representing one-half the major span distance and the other side one-half the minor span distance.

Referring again to the illustration on page 73, draw a rectangle 9 by 16 in. Then draw in a *full* thickness valley rafter. It is very important that the *exact thickness* of the rafter stock be represented, as any variation from the exact thickness of the valley rafter changes the angle of the cuts. Next draw lines at right angles to the center line so that they intersect with that part of the rafter which "touches" the edges of the rectangle. These lines are shown in the opposite corners of the figure.

The length of the center line from the corner of the rectangle to the squared lines is the *run of the cut.* There are *two* runs: the *run of the long cheek cut* and the *run of the short cheek cut.* Note carefully that even though the rectangle is drawn to the $1/12$ scale, the length of the center line from the corner to the squared line, when laying out the cheek cuts, is considered *full length.*

How to Lay Out the Ridge Cut of an Unequal Valley Rafter.—Figure 85 gives a view of each side of the valley rafter, showing both the layout of the ridge cut and the layout of the seat cut. Note that the run of the cut on the side *A* is greater than the run of the cut on the side *B*. To lay out this type cut (not including the shortening):

1. Draw a plumb line near the end of the rafter stock, using "total rise and total run" to get the correct angle of the line.
2. Draw a plan view of the rectangle over which the valley rafter "passes" and develop the two runs of the double cheek cut.
3. Measure at *right angles* to the plumb line on the rafter stock a distance equal to the *long run* of the double cheek cut.
4. Repeat this process on the other side of the rafter stock, laying out the *short run.*

5. Draw a center line down the top edge of the rafter stock.
6. Connect lines as shown. This completes the layout for the double cheek cuts.

If the rectangle has been drawn in relation to its position in the roof and the rafter to its corresponding place in the rectangle, the position of the rafter will determine in which direction to draw the long or short cheek cut.

How to Lay Out the Seat Cut of the Valley Rafter.—(Without Cornice.) The double hollow cheek cut of the seat cut of the valley rafter is laid out in the same manner as the ridge cut. The height of the *level* line of the seat cut is measured on the line length line and is made the same as the common rafter at that point. Observe that where no cornice is involved the long cheek cut of both ridge and seat cut will be on the *same* side of the valley rafter.

Shortening of the Valley Rafter for the Ridge.—(Figure 86.) In the shortening process of the ridge cut of the valley rafter

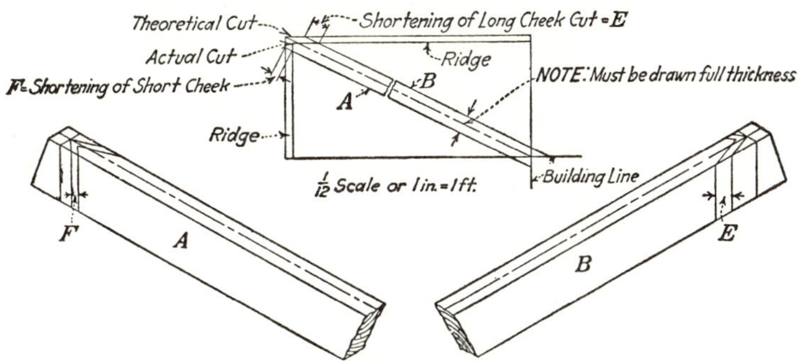

Fig. 86.

each angle of the double cheek cut is shortened a *different* amount. The plan view in Fig. 86 shows the exact amount very clearly. The long cheek cut is shortened the distance labeled *E* while the shortening of the cheek cut is equal to the amount labeled *F*.

Theoretically, the shortening is always done on the center line. Note, however, that the lines that "tie" the theoretical cut with the actual cut are parallel and of equal length. In actual practice, therefore, the length of the line between the theoretical and actual length is the amount to shorten each part of the double cheek cut.

To shorten the ridge cut, then:

1. Measure on the plan view the actual shortening required.
2. Measure the required amount at *right angles* to the plumb cut on *each* side of the rafter stock. The position of the valley rafter will determine which side is to be shortened the lesser amount. Rafter *A* in the drawing shows the side of the rafter that has the lesser shortening while the greater shortening is shown on the opposite side at *B*.

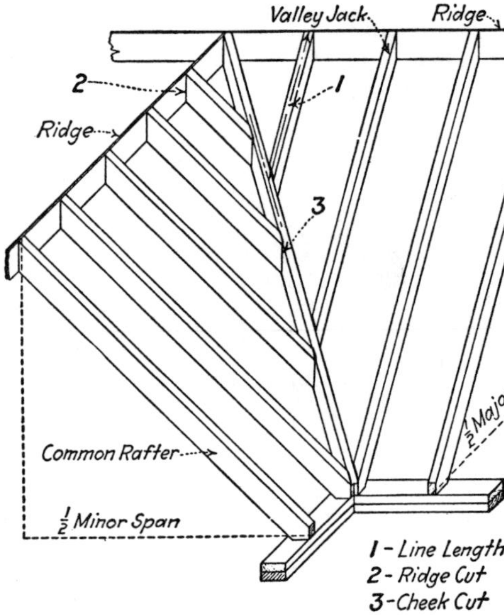

FIG. 87.

3. After the second plumb line has been drawn, use a T-bevel and mark the angle of the cheek cut parallel to the cheek cut lines as first laid out.

If the layout work is done correctly, the meeting point of the two cheek cut lines will *not* pass through the center of the rafter stock. This is shown in both the plan view and the side view of the rafter.

The Unequal Pitch Valley Jack Rafter.—(Figure 87.) The unequal pitch valley jack rafter, when cut and ready to be nailed into the roof, is identical in *shape* with any valley jack, having in common with these rafters a ridge cut and a cheek cut. The order of procedure is first to determine the line length; second, the ridge cut, and third, the cheek cut. As usual, these processes will be discussed in their natural sequence. The drawing shows how the valley jack appears when framed in the roof.

How to Find the Line length of the Unequal Pitch Valley Jack Rafter.—(Figure 88.) The rectangle, drawn to $\frac{1}{12}$ scale and representing each of the two spans of a building, is again shown in this drawing. Within it lines have been drawn at

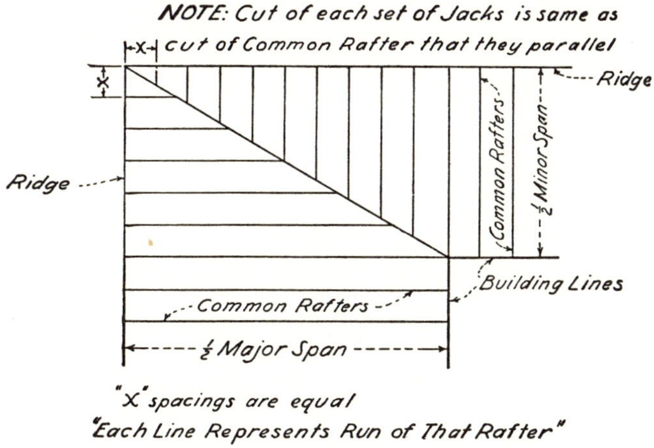

FIG. 88.

right angles to the ridge corresponding to the position of the jack rafters. These lines naturally are also $\frac{1}{12}$ scale and represent the *run* of each jack rafter.

The Cut of the Valley Jack.—The rise and run cut of each set of valley jacks is always the same as that of the common rafter to which *they are parallel.* If they parallel the common rafter of the regular pitch roof, the cut of the jack is the same as the cut of that common rafter. Likewise, the cut of the other set of jacks will be the same as the cut of the minor span which is based on "total rise and total run."

Layout Methods for the Valley Jacks.—When the *run and cut* of the valley jack is known, use the step-off method in the regular manner for each jack.

For the regular pitch roof it is a simple matter to shift the square as many times as the total run is greater than the unit of run (12 in.). The unequal pitch jacks, however, are not so easily disposed of.

To illustrate:

Assume the total run for a given jack is 9 ft. 4 in. and that the cut for that rafter is 10¼-in. rise and 8½-in. run. As 8½ in. does not divide evenly into 9 ft. 4 in., some difficulty would be encountered if the step-off method were used. The problem is simplified, if the unequal pitch cut is changed from an *odd to an even cut* based on the 12-in. run (see next paragraph).

How to Change an Odd Cut into an Even Cut with a Unit of Run of 12 In.—(Figure 89.) Part *A* of Fig. 89 shows the side

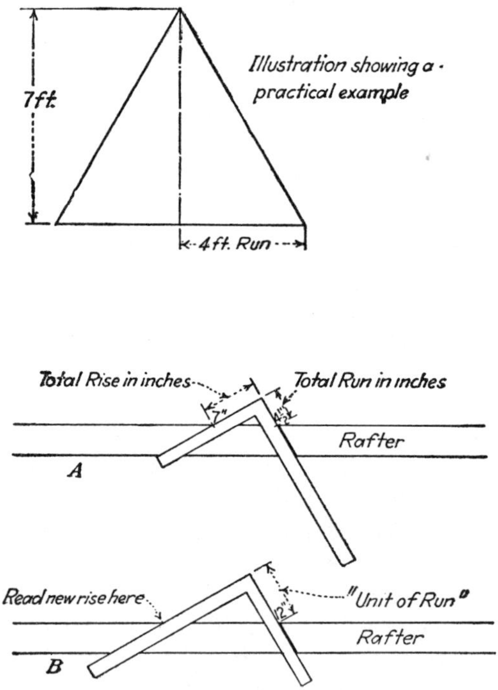

FIG. 89.

view of a piece of rafter stock, with the square applied, using 7 and 4 as the cut (7 ft. as the total rise and 4 ft. as the total run). A mark on the blade (or run side of the square) becomes a level line when the rafter stock is raised to its slant position.

Part *B* of the same drawing is a representation of the same rafter. This time, however, the square is so applied that the figure 12 becomes the "run number" on the blade of the square. The drawing shows the shifted position of the square with its blade *exactly parallel* to the line made when the 7 and 4 cut was laid out.

Now "read" the new rise on the tongue of the square which is 21 in. This number and 12 make the cut that will give an angle similar to the one laid out with 7 and 4 cut. With the new cut, it is a simple matter to use the step-off method as usual, shifting the square as many times as the rafter has total feet of run.

Unequal Pitch Valley Jack Ridge Cut.—(Includes Shortening, Fig. 90.) The ridge cut of the unequal pitch valley jack is

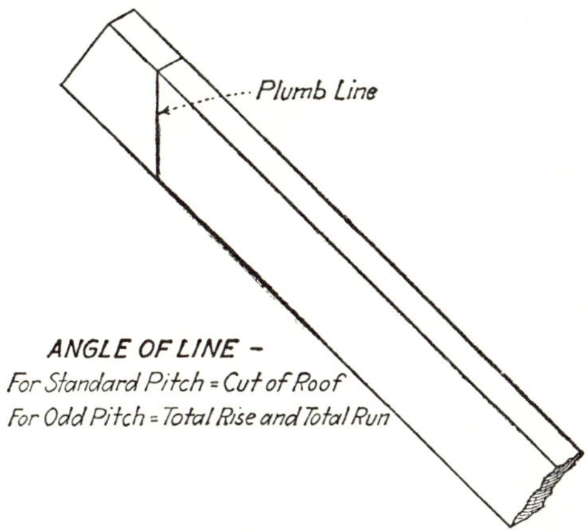

Fig. 90.

identical in shape with that of the ordinary valley jack. For the angle of cut, use the cut of the regular pitch roof for the rafters that belong to that section of the roof. For the unequal pitch set, use *total rise and total run.*

Shortening.—The shortening presents no new problem. Simply measure at *right angles* to the plumb line of the ridge cut, a distance equal to one-half of the thickness of the ridge stock.

Cheek Cut of the Unequal Pitch Valley Jack.—(Shortening not Included, Fig. 91.) The cheek cut of the unequal pitch valley jack is a single cheek cut. The layout work is similar in one respect to the layout process of a cheek cut for a regular pitch roof as the measurement is taken at *right angles* to the

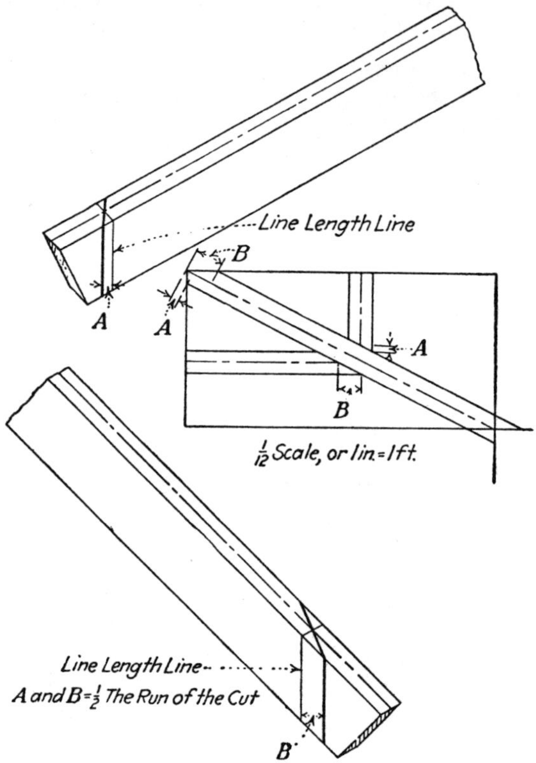

FIG. 91.

plumb cut line. The distance, however, is determined by the shape and size of the rectangle formed by the two spans of the roof.

In the drawing, it will be noticed that the cheek cuts for each set of jacks are *not* at the same angle. In addition to the difference in the shape of the cheek cuts, if the rectangle as drawn is changed in size even a very small amount, the angles of the cheek cut in the plan view are also immediately changed. The problem, however, is simplified by drawing a 1/12-scale rectangle representing the exact measurements of the roof to be framed, as this plan view will show the *exact run of each cheek cut of the valley jacks.* In the layout process, the work is easier if one-half of the run of the cut is used instead of the full run. This distance can be determined from the plan view of the drawing.

Now that the method of finding the run of the cut of the unequal pitch jack has been developed, attention should be called to the fact that one-half the run of the cut of the long jack is *exactly* as much as the long cut of the valley rafter and, likewise, one-half the short cut of the jack is equal to the short cut of the valley. Therefore, when the runs of the valley rafter cuts are known, these same runs can also be used for the jack rafters. In practice, no jack rafter cuts need be developed in the plan view. Simply find the valley rafter cuts and apply them to the jacks.

How to Lay Out an Unequal Pitch Cheek Cut.

1. Draw a rectangle to 1/12 scale, each side being equal to one-half of one of the spans.
2. Draw lines to represent the *exact* thickness of one jack rafter in each set of jacks.
3. Draw lines at *right angles* to the jack rafter lines as shown at *A* and *B*. The distance between these lines equals one-half *the run of the cut.*
4. Measure at *right angles* to the plumb cut line on the rafter stock a distance equal to one-half the run of the cut. Rafters *A* and *B* are jacks showing cheek cuts of different angles.
5. Draw the diagonal line across the top edge of the rafter stock, being sure that it passes through the *center* of the valley rafter stock and the line length line.

Shortening the Valley Jack on the Cheek Cut End.—(Figure 92.) This drawing is a continuation of the plan view or rectangle shown in Fig. 91. In Fig. 92, however, the emphasis is placed on the necessary shortening of the valley jacks to allow for the thickness of the valley rafter.

The amount to shorten, as developed in the drawing and shown at letters *G* and *H*, is *different* for each set of jacks. Note also that this distance is the *actual* amount to shorten.

This rectangle has a peculiar function. The center line of the jack rafter represents the *scale run* of that jack and is measured from the outside of the rectangle to the center line of the valley. If the shortening is being determined, *this same center line between the lines that represent the valley rafter* is the actual dis-

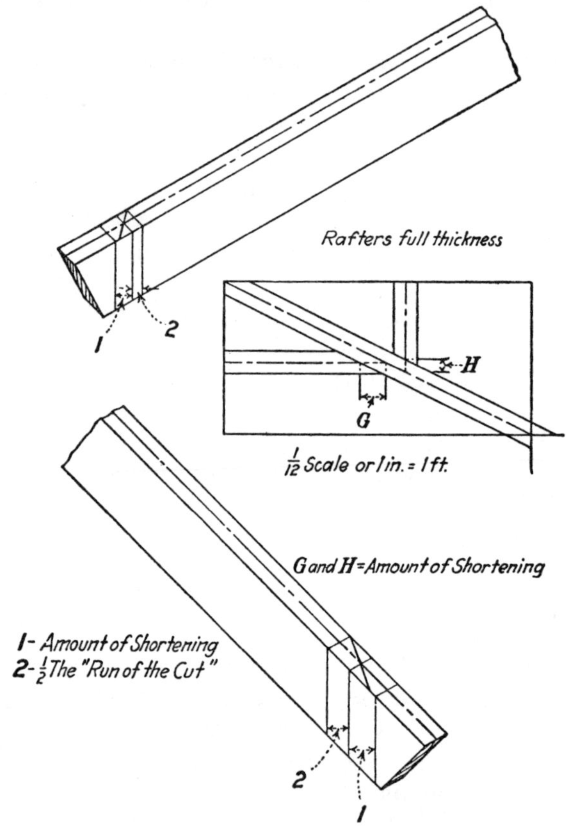

Fig. 92.

tance to shorten the jacks. In other words, there are *two* scales to the same drawing. If run is being figured the scale is $1/12$, but if the amount of shortening is being determined, the scale is *full size* for the part that represents shortening.

The shortening is shown as being allowed for on the cheek cut end as that is where it belongs theoretically. In common practice, however, it will be found expedient if the shortenings for both ridge and valley rafters are added together and allowed for at one time when laying out the ridge cut.

In laying out the valley jack, include its theoretical length, marking both ridge and cheek cuts. Next move the ridge cut line down a distance equal to the two shortenings. The direction of the angle of the cheek cut is of course determined by the position of the rafter in the roof.

Complete Layout Plan for the Unequal Pitch Valley and Jack Rafters.—(Figure 93.) This drawing is a "tying up" in

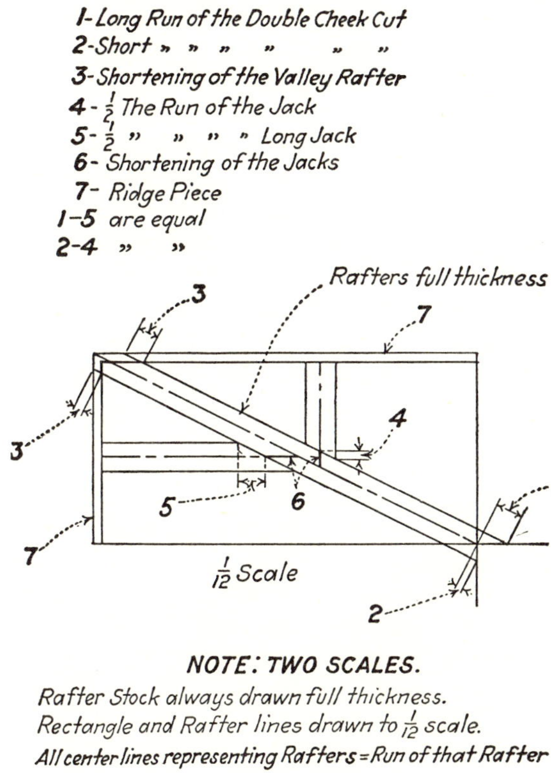

FIG. 93.

one of the several that have preceded it. A study of the foregoing drawings will show that only *one* principle has been developed in each. In actual practice, the making of a separate drawing for each problem involved would entail a great deal of unnecessary work and was done here only to simplify and bring out more clearly the principles involved. Figure 93 illustrates a complete layout for unequal pitch valley and valley jacks. It is important to emphasize again that this drawing has *two* scales and that caution must be exercised in reading the plan to see that the proper valuation is given to the line being measured.

THE UNEQUAL PITCH ROOF WITH A CORNICE

A little study of Fig. 81 will be necessary in order to see why the problem of the unequal pitch roof *with* a cornice differs from one *without* a cornice.

Note the view representing the minor span. The left-side common rafter is shown without an overhang, while the right-side common rafter, with and without an overhang, is drawn showing the rafter in two positions. It is obvious that the *full-line* length rafter is the mate to the left-side common rafter and that the dotted rafter on the right side shows the common rafter with an overhang. It will be seen that the introduction of a cornice changes the pitch of the common rafter which becomes a little flatter as the tail cut "lines up" with the tail cut of the rafters on the main part of the building.

Increase in Total Rise and Total Run.—(Figure 94.) Figure 94 shows a partial end view of the same gable as is illustrated in Fig. 81. The inner triangle represents the roof without a cornice and the outer triangle a roof with a cornice. Distance (1), which equals one-half the minor span, is increased by the *run* of cornice of the main roof (2), while distance (3) is a combination of (1) and (2) and illustrates the *increased run.* Distance (4) represents the total rise of the main roof. This is also increased by the *drop* of the cornice of the main roof (5), and (6) shows the *increased rise* due to the combination of (4) and (5).

Cut of the Common Rafter.—(Cornice Included.) The *cut* of the common rafter for an unequal pitch roof with a cornice is *increased total rise and increased total run.* Otherwise, the rafter is laid out as previously described for the common rafter without a cornice.

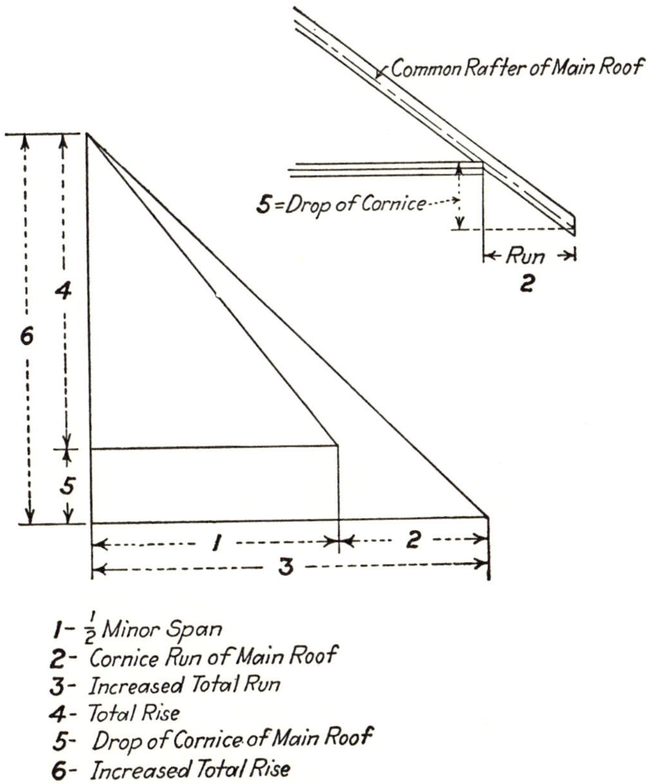

FIG. 94.

How the Cornice Affects the Height of the Top Plate of the Minor Span.—(Figure 95.) When a cornice is added to an unequal pitch roof the eave line of the minor span is lower than the eave line of the main span. This is accounted for by the fact that the roof of the minor span, being steeper, automatically drops the tail cut of the rafters below the line of the tail cut of the main roof rafters.

If the plans of the building having an unequal pitch roof show the minor span gutter line *below* the gutter line of the main span, the common rafter of the minor span presents no problem that has not already been discussed, as all that is necessary is to add a tail cut to the rafter. If, however, the plans show that the gutter lines of the building are to be the same height, the problem arises of planning the roof construction so that the eave line of both the major and minor span are level.

The drawing shows a common rafter in two positions. The lower position represents the common rafter of the minor span

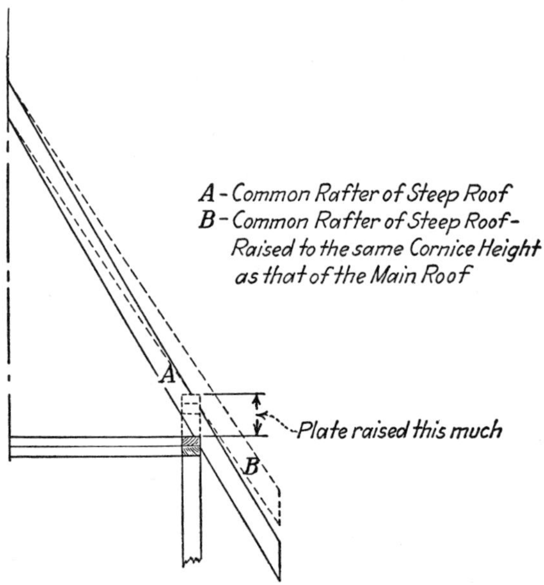

FIG. 95.

continuing from the plate in a regular manner. The upper position shows a common rafter *raised* so that its eave line is at the same height as the cornice of the main part of the roof. Note that the plate line of the building in the second position of the common rafter is now below the rafter stock and consequently must be raised a definite amount to provide a good bearing for the seat cut.

How to Find How Much to Raise the Top Plate of the Minor Span.[1]—(Figure 96.) This drawing is a *comparative pitch*

[1] See page 138 for a short-cut method.

drawing. Note that the two lines representing rafters show the different pitches of the two spans but that the cornice *run* for both spans is the same. The *vertical difference* between the two lines at the building line is the amount that the plate of the minor span must be raised to give the minor span rafters proper bearing.

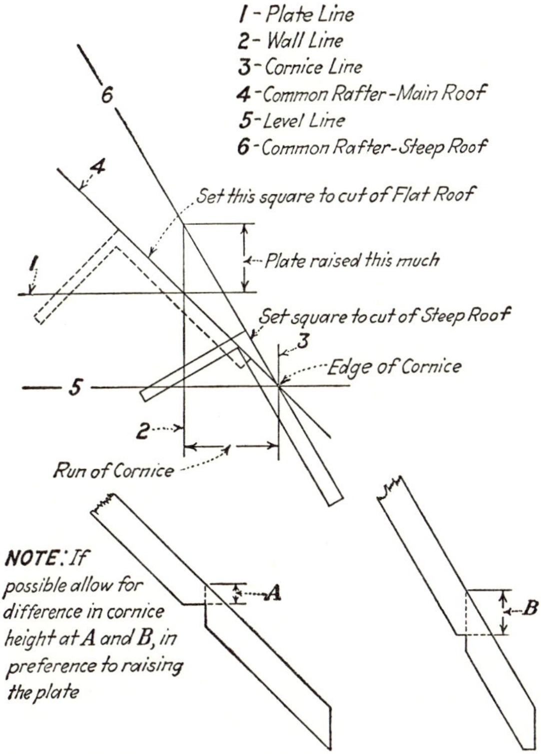

Fig. 96.

How to Make a Comparative Pitch Drawing.—(Figure 96.) To make a comparative pitch drawing proceed as follows:

1. On any flat surface draw a line (1) that represents the level plate line of the building involved, making it full size, if possible, which can be done by drawing it on the sub-floor of the house.

2. At *right angles* to this line draw a line that represents the outside wall of the building (2).

3. Parallel to line (2) draw another line (3), the distance between them to be equal to the *run* of the cornice.

4. Using the steel square, draw a line (4) to represent the common rafter of the regular pitch roof, this slanting line to intersect at the plate line of the building and continue until the cornice edge is reached. The "cut of the roof" will give the proper angle.

5. Draw a line (5) parallel to the plate line, crossing lines (3) and (4) at their point of intersection.

6. Draw another slanting line to represent the common rafter of the unequal pitch roof (6). (Use the steel square to get the correct angle marking on the "run" side.) The cut of this common rafter is *increased total rise and increased total run of the minor span* and the line representing it must be drawn so that its lower end is exactly at the bottom end of the common rafter of the regular pitch. Place steel square as shown in the drawing marking on the "run" side.

7. Measure the *vertical* distance (7) between the two rafters at the *wall line of the building.* This distance is the amount that the top plate must be raised.

If the amount that the top plate must be raised is not too great, make allowance for it in the seat cut of the rafters instead of raising the plate. That is to say, leave as much *more* stock *above* the seat cut of the steeper rafter as the layout drawing shows the plate should be raised. The method to use must be decided before any rafter cutting is done, or the seat cut of the regular rafters may have to be changed.

Unequal Pitch Valley Rafter with a Cornice.—(Figure 97.) By this time it should be apparent that whenever an unequal pitch valley rafter is being considered the run of the valley is always the *diagonal of a rectangle.* When a cornice is added to a building, it is usually customary to make the *cornice run* the same for both parts of the roof irrespective of its pitch. This construction, however, radically affects the run of the "unequal pitch" valley as shown in the drawing, since the run of the valley, with the cornice, is *not* straight as at *A*. It is obvious that it would be impossible to frame a rafter to that run without having to "bend" it. To get around this difficulty, the following procedure is necessary.

Consider the run of the valley from the plan view of the ridge intersection to the outside edge of the cornice as a *straight line* (see *B* in the drawing). Note that the valley rafter does not pass over the building corner, but at a distance away from it. While this may not appear a desirable form of

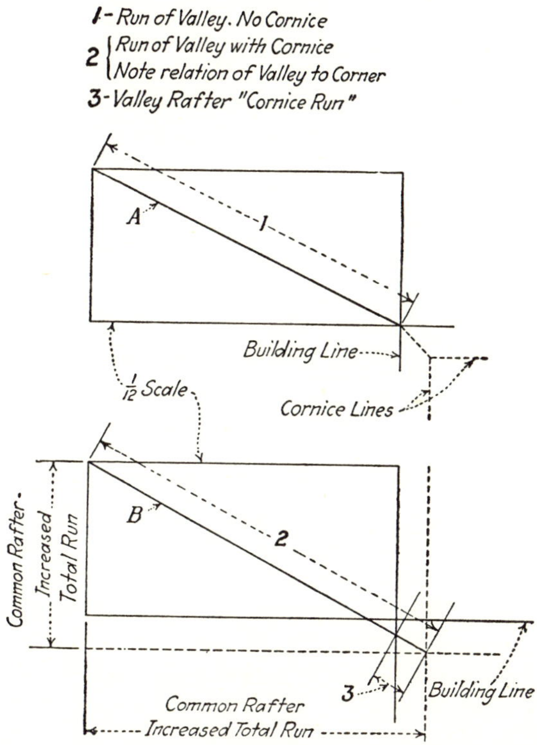

FIG. 97.

construction, it is the only practical solution of the problem and the only method that will give a finished appearance to the cornice.

Cut of the Unequal Pitch Valley Rafter with a Cornice.—(Figure 97.) The cut of the unequal pitch rafter with a cornice will not be total run and total rise as used for the same rafter *without* a cornice. As previously stated, the run of the rafter must be considered to the cornice line of the building. The *cornice run* of the valley will have to be added to the regular run which is the diagonal of the total runs of the two spans. This result is called the *increased run.* This run is most easily found by drawing a plan view of the two spans involved including the run of the cornice. (Figure 85 shows the plan view without a cornice.)

The same problem appears when the total rise is considered. The regular total rise is always the vertical distance that the ridge is above the plate line. To this total rise must be added the *drop of the cornice* inasmuch as the cornice line is below the plate line of the building (see Fig. 81). Therefore the cut of the unequal pitch valley rafter *with a cornice* is always "increased total run and increased total rise."

Line Length of the Unequal Pitch Valley with a Cornice.—The line length of the valley rafter with a cornice is determined by either (1) measuring the bridge distance of the increased total rise and increased total run on the square as represented, or (2) by using increased total rise and increased total run as the cut, and stepping off twelve times.

Tail Cut of the Unequal Pitch Valley Rafter with a Cornice.—(Figure 98.) The tail cut of the unequal pitch rafter with a cornice is developed as outlined for the valley rafter without a cornice (see Fig. 85). The plan view shows the run of each part of the double cheek cut. Measure these distances at *right angles* to the plumb line that represents the line length of the building. For the valley rafter, note that these measurements are taken *below* the line length line.

Seat Cut of the Unequal Pitch Valley Rafter with a Cornice.—(Figure 98.) The seat cut of the valley rafter with a cornice is in the shape of a *single cheek cut* as it does not fit over the inside angle of the building. The plan view will show the exact run of the cut when the rafter is drawn in to net thickness (Fig. 85).

How to Find the Cornice Run of the Unequal Pitch Valley Rafter.—(Figure 97.) The method of locating the seat cut of the unequal pitch valley rafter with a cornice differs from the procedure followed in finding the seat cut of the average rafter in that the first is measured from the *extreme lower end* of the valley rafter stock. The plan view shows how the *cornice run of the valley rafter* is found. When the cornice run has been determined (see (3) on the drawing), use the step-off method and work toward the *top* end of the rafter.

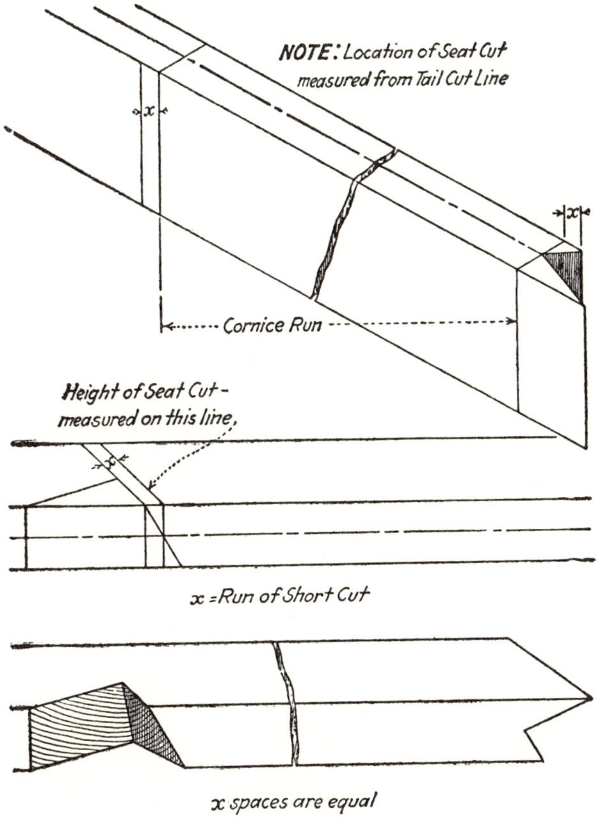

FIG. 98.

How to Locate the Center Line of the Seat Cut.—(Figure 98.)

1. Draw a plan view of the roof to be framed, including the cornice run.
2. Draw a line that represents the run of the valley rafter.
3. Measure the valley cornice run.

4. Step off the cornice run distance.

Note: It will probably be found necessary to change the cut of the valley rafter to one with the 12-in. unit of run (see p. 82).

5. When the proper distance has been stepped off, draw a plumb line. This line locates the center of the seat cut.

How to Lay Off the Seat Cut.—(Figure 98.) The seat cut is laid off similarly to that of any unequal pitch cheek cut.

1. Find one-half the run of the seat cut. This is equal to short run of the valley rafter double cheek cut as developed on the plan view (see (2) on Fig. 93).

2. Measure at *right angles* to the plumb line found in step (5) above a distance equal to the run of the short cut, and make another plumb line at that point. The direction to measure (above or below the building line) is determined by the position of the rafter. If two valleys are required, either direction will do as the valleys must be made in pairs and the second rafter can be cut the reverse direction of the first one. If only one valley is required, the plan view should be made showing the valley rafter in its true position.

To review briefly the layout of the unequal pitch roof:

1. Make a plan view of the rectangle over which the valley rafter "passes."

2. Develop this plan view so that it shows the run of all the rafters, both jacks and valleys.

3. Also develop on the same drawing the run of all the cheek cuts involved.

4. Be very careful in "reading" the drawing to see that the proper valuation is given to each line. That is, if run is being measured, remember that all run lines are drawn to the 1/12 scale. If shortening or run of cuts is being measured, remember that all of these lines are *full* scale.

5. The cut of the unequal pitch rafters is "total rise and total run" except in the case of the common and valley rafter with a cornice when the cut must be "increased total run and increased total rise."

6. When stepping off, if the rafter cut is an "odd" one, change to a run with 12 in. as its base and proceed as usual. Use a sharp pencil, draw the lines accurately, and read the drawing carefully.

CHAPTER VI

SPECIAL PROBLEMS

1. The snub gable.
2. The odd-pitch roof.
3. How to determine the cut of a roof from a blueprint.
4. Common difference in the length of jack rafters for any equal pitch roof.
5. How to lay out a jack rafter cheek cut for any pitch roof jack rafter.
6. The level cheek cut that fits on a roof of the same pitch.
7. The level cheek cut that fits on a roof of a different pitch.
8. Cut of a rafter that "dies" on another roof of the same pitch.
9. The shed dormer roof cut.
10. The shed roof rafter.
11. The gable stud cut.
12. Special ridge cut.
13. Common difference in the length of gable studs.
14. Determining the length of a collar tie.
15 Determining the length of ceiling joists that frame to a rafter.
16. Purline, frieze and sheathing cuts.
17. Cuts of a roof in terms of degrees.

CHAPTER VI

SPECIAL PROBLEMS

There are a number of special problems that indirectly belong to the framing of gable, hip, or intersecting roofs. These odd problems are discussed in a separate chapter.

There is no definite relation between any of the cuts or problems given. Each drawing or group of drawings that illustrates one of the problems and the descriptive matter relating to it is a unit complete in itself. Where the basis of the layout work has been described in previous chapters, the description or principles will not be repeated.

1. THE SNUB GABLE

The snub gable (Fig. 99) is, as its name implies, a type of gable roof that has been "snubbed" at the top. Its architectural purpose is a sufficient explanation for its use.

The size of a snub gable is determined by the height of its plate line; that is, its perpendicular height above the plate line of the main roof. For accurate work it is not always satisfactory, as before pointed out, to scale this measurement from the elevation sheet, as very often the roof is not drawn to its exact mathematical height.

Rule for Finding Mathematical Height.—A study of the plan view of the snub gable, as illustrated in the drawing, will show that as usual the run of the hip rafter is at an angle of 45 deg. to the outside of the building. This at once simplifies the problem, as it can readily be seen that the length or span of the snub gable as shown at (2) is twice the distance X, or the amount that the last full-length common rafter is *in* from the end of the building. To reverse this statement, the last full-length common rafter is always placed *in* from the end of the building a distance equal to *one-half* of the length of the snub-gable plate length.

How to Find the Height of the Snub-gable Plate.—The end view of the gable roof in the figure shows that in the snub-gable type of roof there are *three* spans to be considered. They are:

1. The span of the main roof.
2. The span of the snub gable.

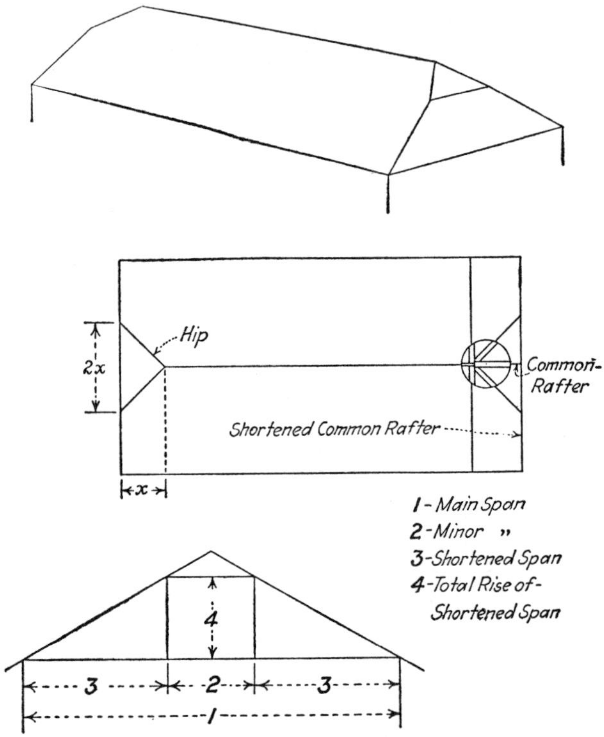

Rule for finding the Plate Height: Pitch times Shortened Span **3** *= Total Rise* **4**

FIG. 99.

3. The *shortened span*, the length of which is the *difference* between the length of the main span and the snub-gable span (sometimes called the minor span).

The height of the plate on which the snub gable is built is determined by the length of the *shortened span.* Use the formula given in Chap. I, "span times pitch equals total rise."

To illustrate:

Assume as the main span of a gable roof a distance of 28 ft. and a minor span of 8 ft. The *shortened span* is then the difference between 28 ft. and 8 ft., or 20 ft. If the roof has a 1/4 pitch, the height of the snub-gable plate above the top plate of the building is 20 ft. × 1/4, or 5 ft.

Note that while this is a *mathematical* height, it is also the *actual* height, as the amount of rafter stock above the seat cut of the common rafter should be the same as on the common rafter of the snub gable. If 2 in. of rafter stock is left above the seat cut of the common rafter of the main roof, 2 in. should also be left above the seat cut of the snub-gable common rafter.

Construction of a Snub Gable.—When the length of the snub-gable plate is known and its height has been determined, frame the level plate in place, bracing it to some convenient part of the building. The common rafters of the main roof can also be raised as the position of the last pair is known to be one-half the length of the snub-gable plate *in* from the end of the building.

The framing of the snub gable, then, resolves itself into an ordinary hip roof problem, the snub gable differing only in its size and position in the roof.

Kinds of Rafters.—(Figure 100.) There are *three* rafters to be framed in a snub gable: the common rafter, the hip rafter, and a new type of rafter called the *shortened common rafter.*

Common Rafters.—The common rafter is cut exactly as any common rafter and has a corresponding run of one-half the minor span. The drawing shows the finished rafter cut to its "mathematical" length. Note the shortening, which equals one-half the thickness of the full-length common rafter stock. Note also the construction within the circle in the plan view (Fig. 99), showing the snub-gable common rafter framed to the *main common rafters.*

The Hip Rafters.—The hip rafters are laid out as usual. Use either the step-off or the slant triangle method (see Chap. III, pp. 25–27). This hip differs in one respect from the ordinary hip in that the tail cut is a *single* and not a double cheek cut. The ridge cut, however, is a regular double cheek cut. The seat cut is best framed by making the "hollow *V*" cut which permits it to fit snugly over the end of the plate stock.

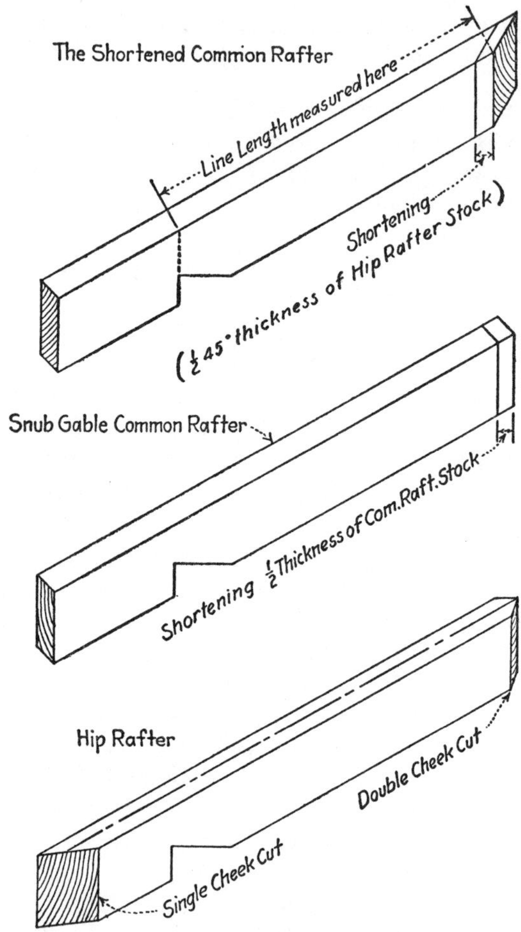

FIG. 100.

The Shortened Common Rafter.—The shortened common rafter when framed looks like a hip jack rafter, and in one sense this is what it really is. The *line length* of this rafter differs from any other rafter in a regular roof in that it is measured to the *short point of the cheek cut.* This is because the short point is the outside face of the rafter which must line up with the outside face of the building.

Operations in Logical Order for Framing a Snub Gable.—To review briefly the problem just discussed:

1. Determine the length of the minor or snub-gable span (usually from the blue print).
2. Using this as a basis, find the shortened span distance.

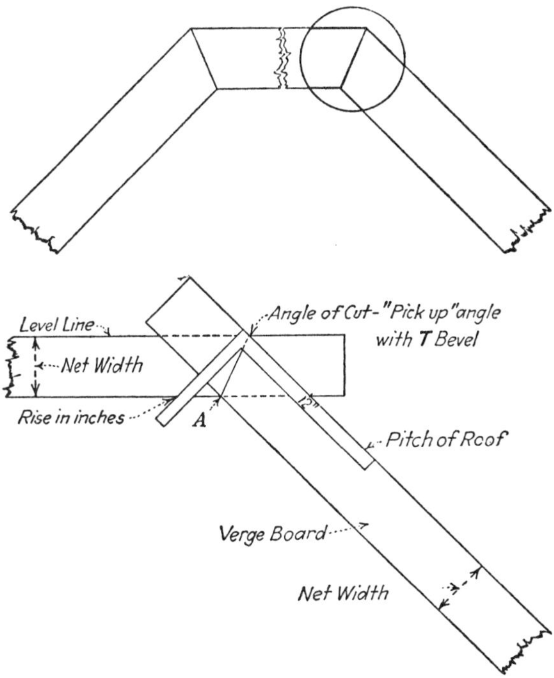

Fig. 101.

3. Multiplying the shortened span distance by the pitch, find the exact height of the snub-gable plate.
4. Cut and erect to place this plate stock.
5. Raise and frame the main rafters of the roof. (Of course, this step can be carried out either before or after the snub-gable problem is worked out.)
6. Cut and frame in place the necessary snub-gable rafters.

The Verge Board Cut for the Snub Gable.—(Figure 101.) If a verge board is used on a snub gable, the angle of the cut for the intersection must be determined. The drawing illustrates that cut. The layout procedure is:

1. On a flat surface draw a line to represent the level plate of the snub gable.
2. Draw a slanting line to the "cut of the roof," intersecting the level line.
3. Draw a line *parallel* to each of these lines. Distance between is the *exact width* of the verge board stock.
4. Connect the outside angle of the lines with the inside angle as shown at *A*. This line will be the shape of the verge board at the intersection joint.

Using a T-bevel, "pick up" the angle thus determined. Then transfer or lay out the angle on the end of the verge board stock.

2. THE ODD-PITCH ROOF

An odd-pitch roof is one whose total rise is *not* an even fractional part of the span (see Fig. 102).

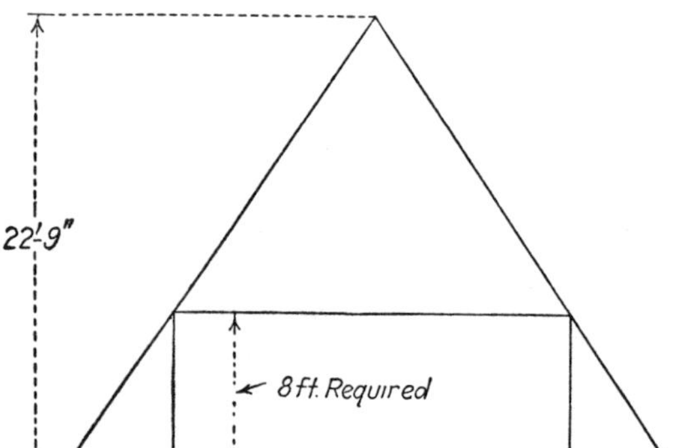

Fig. 102.

To illustrate:

Due to the fact that in this example the plans call for an 8-ft. ceiling in the second story, the roof, as shown, developed into a total rise of 22 ft. 9 in. The span is 30 ft. The pitch which is the relation of 22 ft. 9 in. to 30 ft. equals $^{91}/_{120}$

The cut for a pitch such as this would be hard to determine in terms of the usual unit of run (12 in.). Therefore, instead of attempting to do so, employ the quicker method based on *total rise and total run.* If these figures go beyond the limits of the square, divide them both by 2 and double the diagonal distance. Since, with the gages set on the square to total rise and total run, the scale would be 1 in. to 1 ft. or $\frac{1}{12}$ scale, use the step-off method and repeat twelve times. Twelve steps are always taken when total rise and total run are used, because the gages as set on the square measure one-twelfth the actual run of that rafter.

Stated as a rule, for an odd-pitch roof, use total rise and total run and step-off twelve times.

3. HOW TO DETERMINE THE CUT OF A ROOF FROM A BLUE PRINT (FIGURE 103)

Occasionally a set of plans for a building are drawn without stating either the cut of the roof or the pitch. While the total rise of the gable might be scaled in order to find the cut, this is

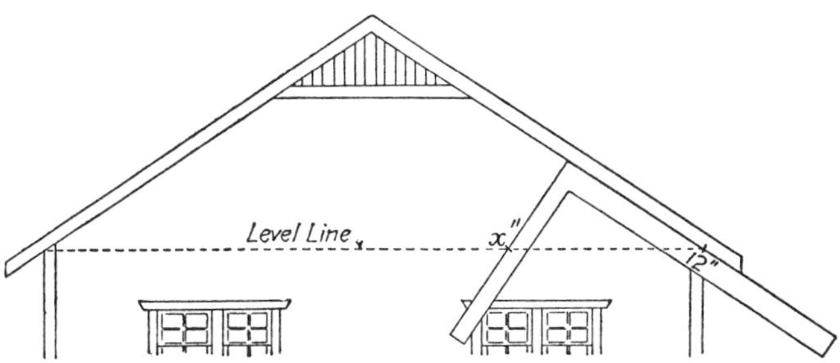

FIG. 103.

not as a rule satisfactory, as the roof of an ordinary plan is not always drawn to its exact mathematical height.

Assuming the slant of the roof to be correct, the cut can easily be found as illustrated in Fig. 103. The procedure is as follows:

1. If the ceiling height of the building is not shown, draw a pencil line on the plan. This line must be level.

2. Lay the square on the plans, letting the blade be exactly parallel with the rafter line and the figure 12 on the level line.

3. "Read" the number on the tongue of the square where it crosses the level line. This number and 12 will be the cut of the roof.

4. COMMON DIFFERENCE IN THE LENGTH OF JACK RAFTERS FOR ANY EQUAL OR ODD PITCH ROOF (FIGURE 104)

The principles underlying jack rafter lengths and cuts have already been developed and discussed in Sec. 2 of Chap. III; therefore no further description is necessary.

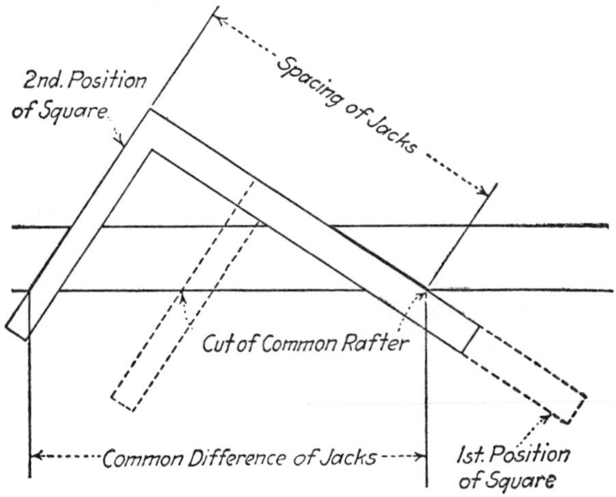

FIG. 104.

The following is given as a short-cut method for finding the *common difference* in jack rafters for any equal or odd pitch roof.

As the "run of a jack is always equal to its distance from the corner" (Fig. 44), likewise the line length of a jack for a given run can be found by stepping off the required run and measuring the bridge distance of the rafter so laid off.

The common difference in jack rafters is *always* the same as the length of a rafter that has as much run as the rafters are spaced.

Therefore:

1. Lay off a section of a rafter by using the cut of the roof and marking on the run side of the square.

2. "Slip" the square so that the blade represents the distance equal to the required spacing of the jacks, keeping the blade even with the line drawn.

3. With the square in this new position, draw a line along the tongue or rise edge of the square.

4. Measure the bridge distance between the rise and run marks. This will be the common difference in the jacks.

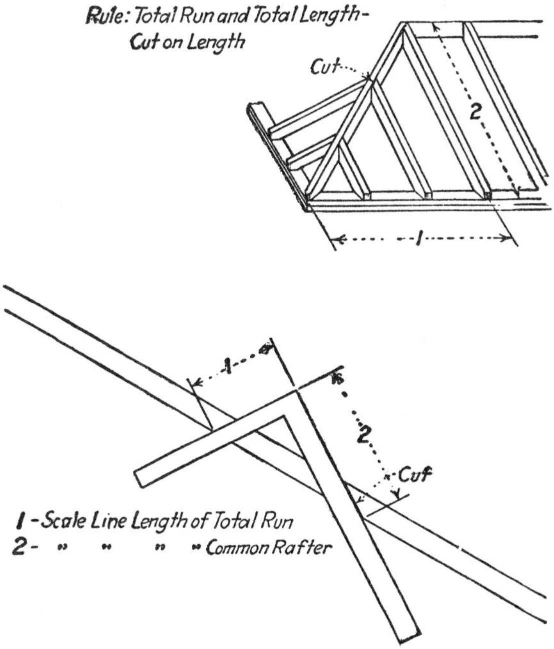

FIG. 105.

This method will always apply to the rafters of any equal or odd pitch roof as long as the run line is drawn to the cut of the roof.

5. HOW TO LAY OUT A JACK RAFTER CHEEK CUT FOR ANY PITCH ROOF[1] (FIGURE 105)

In all of the processes described in the previous chapters for the laying out of a cheek cut, attention has been focused on doing the actual layout work based on the plan view. By so

[1] See pages 129 and 150.

doing, particularly in the equal pitch roof, the workman is not burdened with the use of figures representing the actual length of the common rafters of the roof involved. Likewise, in the unequal pitch rafter problem the "run of the cut" was developed in the plan view.

There is, however, a method by which the angle of the cheek cut can be laid off by the use of the steel square if the carpenter so desires. This method is based on the slant triangle (see Fig. 28). Three rules can be used.

Rule 1. Use 12 in. (unit of run) and length of common rafter per foot of run (diagonal of the cut) and cut on length.

To illustrate: A 1/3 pitch roof has an 8 and 12 cut. The length of rafter per foot of run for this cut is 14 3/8 in. Use 12 in. and 14 3/8 in. and cut on the latter.[1]

Rule 2. Represent one-half of the span distance on the tongue of the square and the line length of the common rafter on the blade and cut *on length.* To abbreviate the rule, "Total run and total length, cut on length."

It is obvious, of course, that the measurements used for the total run and the line length of the common rafter must be taken from the roof being framed.

Rule 3. For the unequal pitch, the total run of the main span is used with the common-rafter length of the *minor* span and, vice versa, the total run of the minor span is used with the common rafter length of the *main span.*

6. THE LEVEL CHEEK CUT THAT FITS ON A ROOF OF THE SAME PITCH (FIGURE 106)

On some buildings it is advisable to build a dormer gable after the main roof has been framed and sheathed. The advantage in this type of construction lies in the fact that valley rafters are eliminated and that the common rafters can be used full length, making the supporting roof not only stronger but much simpler to frame.

The bottom cut of the rafters that rest on top of the main roof is called the *level cheek cut* and is laid out as follows:

[1] Table II on page 125 gives the length of rafter per foot of run for a number of standard pitches.

1. Using the cut of the roof, mark off a level line on a piece of rafter stock as at (1).

2. Measure *on this line* from the top edge of the rafter a distance equal to the *thickness of the rafter stock*, as at (2).

3. Mark a plumb line at this point, carrying this line to the top edge of the rafter stock and "squaring" it across the top (continuation of line (3)).

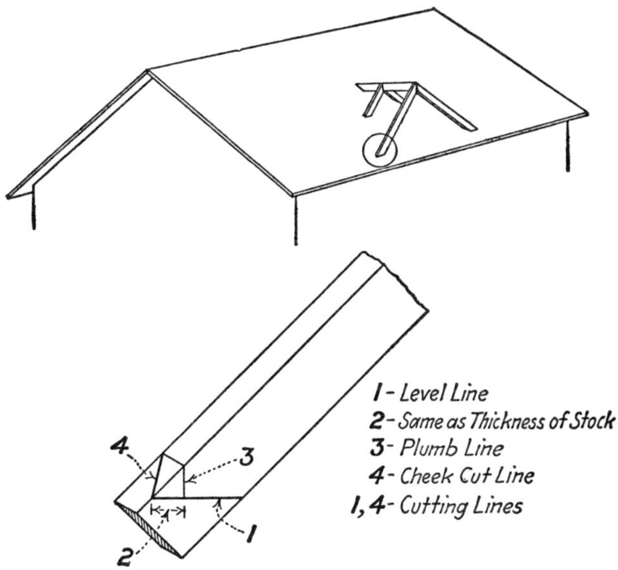

Fig. 106.

4. Connect the ends of lines (1) and (3) as at (4). Sawing on (1) and (4) gives the level cheek cut. If correctly done, the face of the rafters when placed should be plumb, and the bottom cut should exactly fit the slant of the roof.

7. THE LEVEL CHEEK CUT THAT FITS ON A ROOF OF A DIFFERENT PITCH (FIGURE 107)

The shape of a level cheek cut that fits on a roof of a different pitch is identical with the cheek cut just described in the preceding section.

The process of laying out the cut, however, as the drawing shows, is entirely different. To lay out this cut, proceed as follows:

1. Set the square gages to the cut of the steep roof and place as shown by the first position of the steel square (represented by the dotted lines in Fig. 107).

2. Mark a sharp line on the run side of the square.

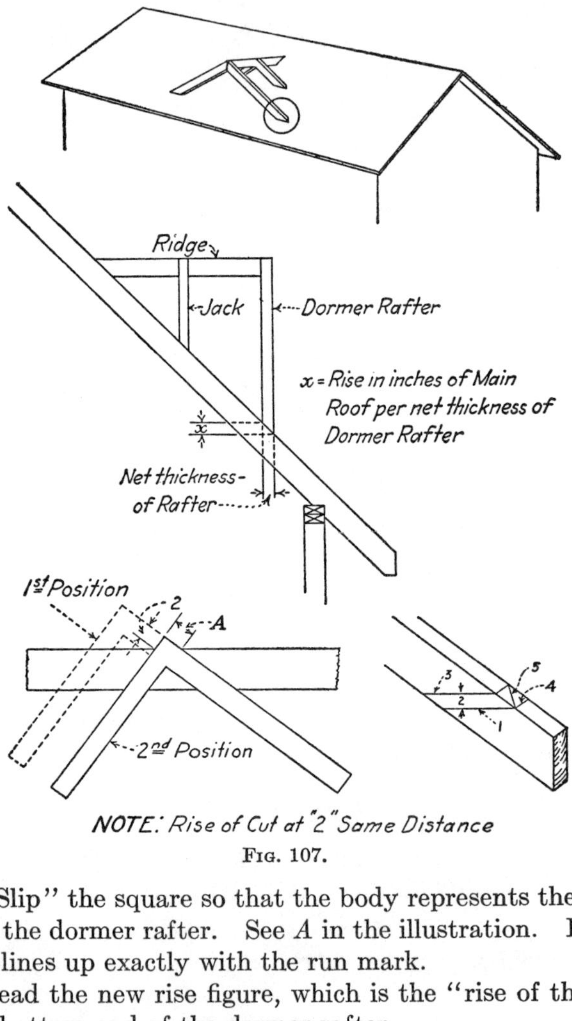

FIG. 107.

3. "Slip" the square so that the body represents the thickness of the dormer rafter. See *A* in the illustration. Be sure square lines up exactly with the run mark.

4. Read the new rise figure, which is the "rise of the cut" on the bottom end of the dormer rafter.

5. Lay out a level line on the dormer rafter as at 1 in the illustration.

6. Lay out a second level line above the first level line as at 3. Distance between the two level lines *vertically* (right angles to the level line) is the distance found in step 4 above, that is, the "rise of the cut."

7. Make 90-deg. lines across the top of the rafter as at 4.

8. Connect with a diagonal line, which will be the angle cut of the dormer rafter as at 5.

8. CUT OF A RAFTER THAT "DIES" ON ANOTHER ROOF OF THE SAME PITCH (FIGURE 108)

Occasionally in roof framing, one gable roof will "die" on top of another gable roof, requiring a special cut as shown in the drawing.

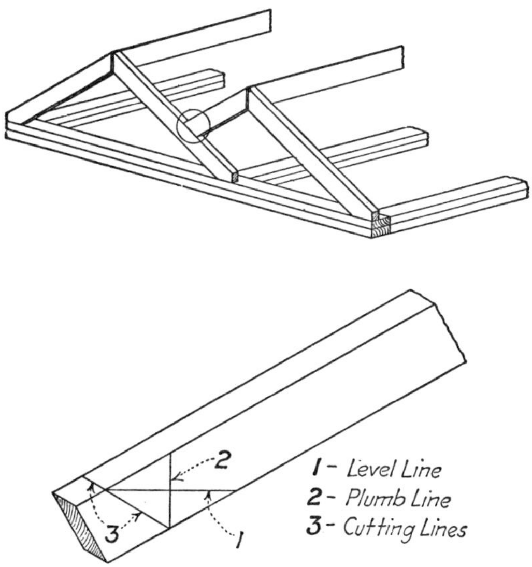

FIG. 108.

To make this cut:

1. Draw a level line on a piece of rafter stock using the cut of the roof as the angle.

2. On this level line mark a point at its center.

3. Draw a plumb line through the center point.

4. Connect the ends of lines (1) and (2) with line (3). Line (3) is the cutting line.

9. THE SHED DORMER-ROOF CUT (FIGURE 109)

The dormer-roof cut is required on the top end of a slanting rafter in order that it may fit on another roof of a different pitch. This is illustrated in the drawing. The layout work is based on the level run of both roofs. In order to explain clearly how to find this cut, a practical example will be given.

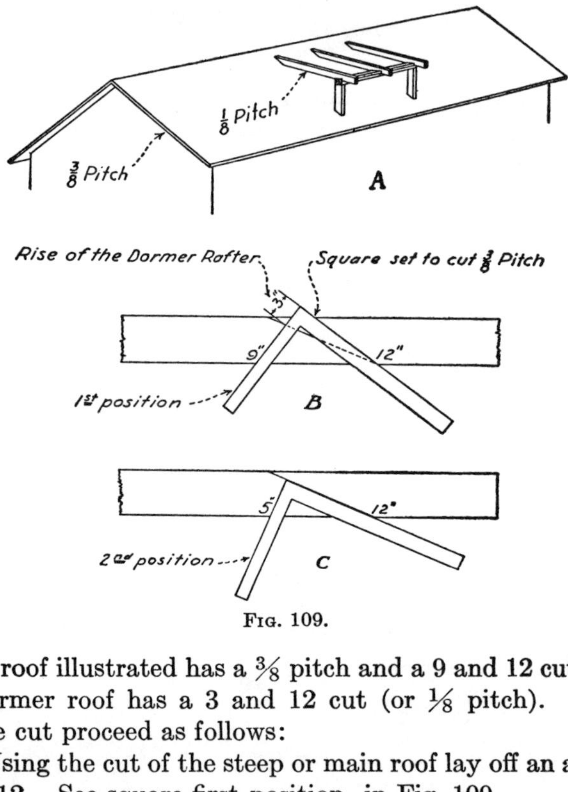

Fig. 109.

The roof illustrated has a ⅜ pitch and a 9 and 12 cut, while the dormer roof has a 3 and 12 cut (or ⅛ pitch). To lay out the cut proceed as follows:

1. Using the cut of the steep or main roof lay off an angle of 9 and 12. See square first position in Fig. 109.

2. While the square *remains* in this position mark the rise of the dormer or shed roof, in this problem 3 in., measuring from the heel or corner of the square.

3. Connect this point with the figure 12 on the body of the square. This will be the cut for the top end of the dormer rafter.

4. "Slip" the square to 12 in. on this developed line and read the rise cut for the dormer rafter on the tongue of the square. In the problem illustrated this will be 5 in. which, with 12, gives a cut of 5 and 12. See square second position in Fig. 109.

10. THE SHED-ROOF RAFTER (FIGURE 110)

The shed-roof rafter is different from the common rafter in that it has *two* seat cuts. The layout process, generally speaking, is the same as for any rafter. Note on the drawing that

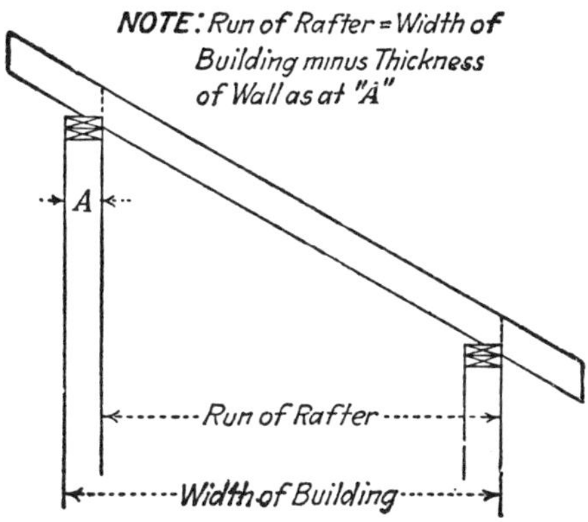

FIG. 110.

run is *not* the full width of the building but the full width *less* the thickness of one of the walls. The dotted lines in the illustration are parallel to each other and are in the same relative position to the plate stock. Therefore:

1. Deduct from the width of the building the thickness of one of the walls.
2. Using the remainder as the run, step off the rafter, laying out *two* seat cuts that are exactly alike.
3. Add as much overhang as is desired.

11. THE GABLE STUD CUT (FIGURE 111)

The top end cut of the gable stud that fits against the under side of the rafter is laid out by using the "cut of the roof" and marking on the *rise* side. The drawing shows the square in position.

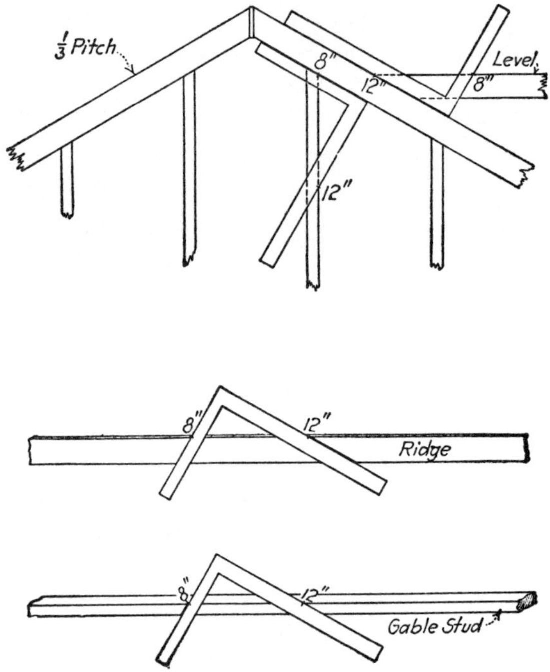

Fig. 111.

12. SPECIAL RIDGE CUT (FIGURE 111)

The special ridge cut as shown in the same drawing is made by using the cut of the roof and marking on the *run side* of the square. Note the square in position.

13. COMMON DIFFERENCE IN THE LENGTH OF GABLE STUDS[1] (FIGURE 112)

The job of "cutting in" gable studs is as a rule a very inconvenient one due to the fact that the work is generally done from on top of the ceiling joists. To eliminate the inconvenience, the studs can be cut on the ground, provided the *common difference* in the length of each gable stud is known. This common difference is found as follows: multiply the *cut* of the roof, expressed as a fraction, by the spacing of the studs. The

[1] See short cut, page 136.

result is the common difference in the length of the gable studs. To illustrate:

Assume a cut of 6 in. and 12 in. and a stud spacing of 16 in. The common difference in the length of the gable studs is 16 in. × 6/12 or 8 in. Expressed as a formula, Stud spacing × Cut of the roof = Common difference.

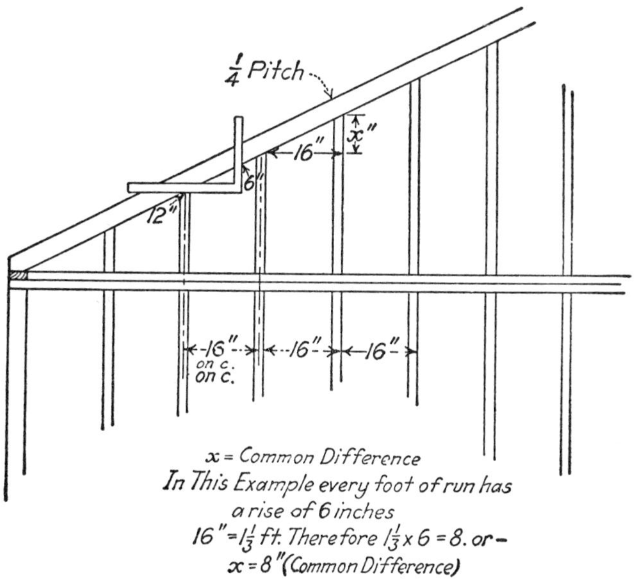

FIG. 112.

The quickest method of finding the length of the *first* gable stud is to place the stock in position and mark the length. The first stud to be cut is usually the longest one. The length of the other studs can be found by marking each the "common difference" shorter than the previous one. It should be further noted that unless the first gable stud is placed exactly in the center of the gable, *two* patterns are required.

14. DETERMINING THE LENGTH OF A COLLAR TIE[1] (FIGURE 113)

A collar tie is a piece of stock (usually 1 × 4 or 1 × 6) fastened to one pair of rafters in a horizontal position at some desired location, between the ceiling joist and the ridge of

[1] See short cut, page 136.

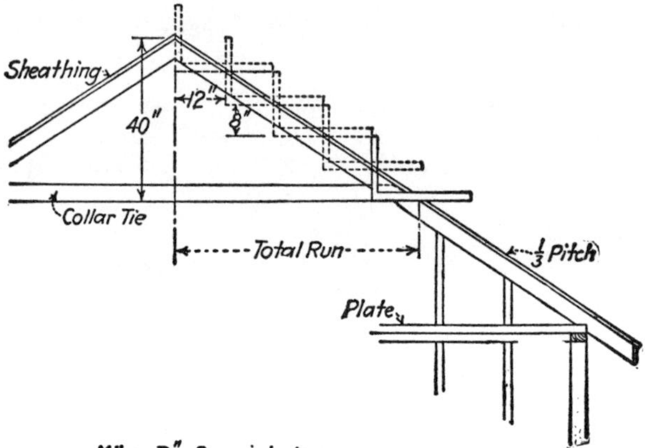

When D" = Drop in inches
" R" per Run = Rise in inches per foot of run
" L = Length Twice the total Run

$$2\left(\frac{D''}{R''\text{ per }'\text{ Run}}\right) = L$$

In this problem, 8 and 12 = cut
40" = Drop of Collar Tie (from peak of roof to under side of collar tie)

$$\frac{40}{8} = 5. \quad 2 \times 5 = 10 \quad \text{Ans. } 10 \text{ ft.}$$

NOTE: For answer in inches multiply by 12

FIG. 113.

the roof. It serves to keep the building from "spreading." Therefore, the lower the tie is from the ridge, the better it fulfils its purpose. Also, the lower it is, the longer it is. Its length is found as follows:

Divide the "drop" (in inches) by the *rise of the roof per foot of run.* This will equal one-half the length of the tie. Doubling this result will give the actual length of the collar tie. The rule can be expressed briefly as follows:

$$\frac{\text{Drop}}{\text{Rise}} \times 2 = \text{length of tie}$$

To illustrate:

Assume a ⅓-pitch roof. It is desired to put a collar tie 3 ft. 4 in. down from the ridge. What is the length of the collar tie?

Solution: A ⅓-pitch roof has an 8 and 12 cut. Converting the drop into inches we have: 40/8 which is 5. 5 × 2 equals 10, length of the tie. The length can be either the top-edge length or the bottom-edge length of the collar tie. The angle of the end cut is made to the "cut of the roof," marking on the 12-in. side of the square.

Note: Dividing inches by inches will not give feet as a result. What is actually done is to find how many *steps* of 8 in. each are necessary to lay off 40 in. of drop. For every 8 in. step down, there will be 12 in. of run. Hence five steps mean 5 ft. of run. This only represents one-half of the roof. Multiplying by 2 allows for the other half.

15. DETERMINING THE LENGTH OF CEILING JOISTS THAT FRAME TO A RAFTER (FIGURE 114)

Occasionally it is desired to increase the ceiling height of a room in the second story of a house by "raising" the ceiling

Cut of Roof = 9" and 12"

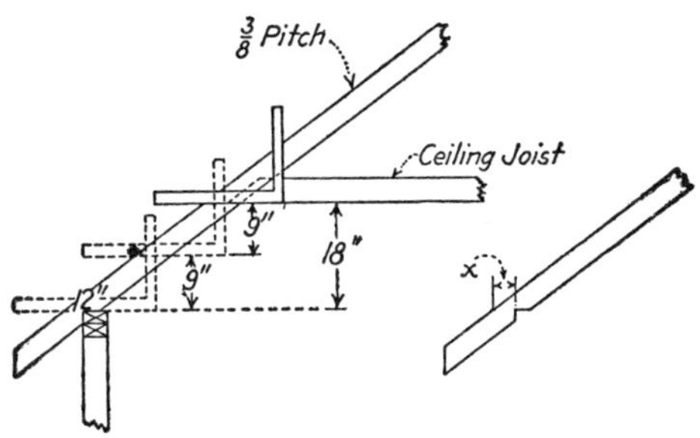

Rule. To find how much shorter than the span to make the ceiling joist–

$$\frac{\text{Vertical rise in inches of Ceiling Joist above plate}}{= \text{Rise in inches per foot of run}} \text{ divided by–}$$

In this example $\frac{18''}{9''} = 2 \times$ *unit of run in inches (12") = 24" one end.*

For both ends 2 × 24" = 48" or 4 ft. – Reduced to a short Formula–

$$12\left(\frac{VR''}{R''\text{per ft. Run}}\right)2$$

NOTE: *For close measuring twice the level distance from corner of seat cut to outer edge of rafter must be added. See "x"*

FIG. 114.

joists above the plate line of the building. The joists are then shorter than the length of the span of the building as they will have to be framed to the rafters. The length is determined by finding *how much shorter* the ceiling joist will be than the span of the building. This is done as follows:

Divide the *vertical rise* of the joist above the plate by the *rise of the roof per foot of run.* This determines the amount to *shorten* the joist at one end. A similar amount must be deducted at the opposite end.

To illustrate:

The pitch of a roof is 3/8. The span is 24 ft. It is desired to raise the joists 18 in. How long will they be?

Solution: A 3/8 pitch has a 9 and 12 cut. 18 in. divided by 9 equals 2. 2 × 2 equals 4 ft. or the amount that the joists must be shortened. 24 ft. minus 4 ft. equals 20 ft., length of ceiling joists.

Note: The 9 in. represents steps and not inches. Dividing 18 in. by 9 equals two steps. As the run of each step is 12 in. the run for the two steps would be 24 in., or 2 ft., which is one-half of the shortening.

The end cuts are laid out to the cut of the roof, marking on the 12-in. side of the square.

16. PURLINE, FRIEZE, AND SHEATHING CUTS (FIGURE 115)

Purline and Frieze Cuts.—A purline or frieze cut is the cut on the end of a piece of purline or frieze stock that permits it to fit to a hip rafter while at *right angles* to the plane of the roof.

The cut has two parts: (1) the *face* or side cut on the stock, and (2) the *miter* or top cut on the stock. *Each* cut requires a separate layout and can be made with the framing square, using a combination of either: the unit of run (12 in.) and the length of common rafter per foot of run (diagonal of the cut), or the rise in inches and the diagonal of the cut.

For the face or side cut use "rise and length" cut on *rise.*

For the miter or top cut use "run and length" cut on *run.*

To illustrate:

Assume a 1/3-pitch roof. The cut for a 1/3-pitch roof is 8 and 12. The diagonal of the cut is 14 3/8 in. Therefore, for the miter cut, use 12 in. and 14 3/8 in. and mark on the 12-in. side of the square. For the face cut use 8 and 14 3/8 and cut on the 8-in. side of the square.

Sheathing Cut.—The cuts to permit the sheathing to fit to a hip rafter are similar to the purline and frieze cuts. The face cut is made on the *edge* of the board.

For the miter or edge cut, use "run and length" cut on *run*. For the face or top cut, use "rise and length" cut on *rise*.

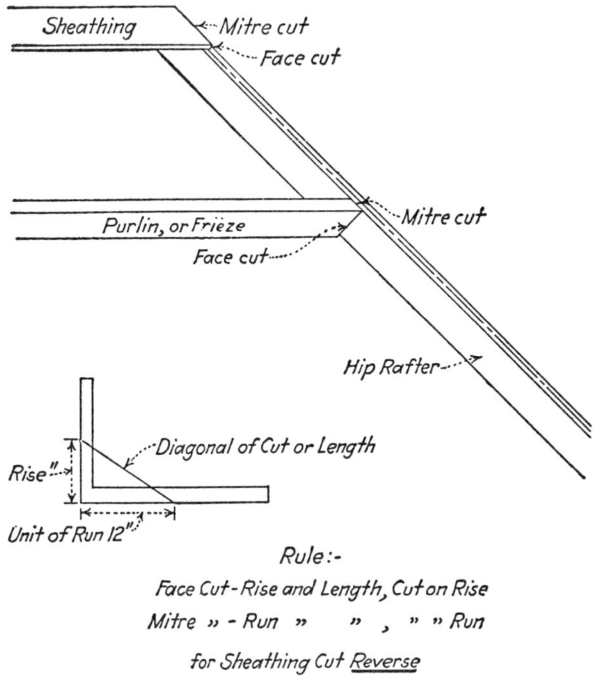

FIG. 115.

17. CUTS OF A ROOF IN TERMS OF DEGREES

It is the practice of some architects to give the angle of slope for a roof in *degrees* instead of pitch or cut. Before the layout work for the rafters can be done, the cut must be found: In the following table the cuts of a roof from ½ in. and 12 in. to 24 in. and 12 in. are given, each cut being ½ in. more than the one that precedes it. Opposite each cut is given the corresponding degree for that cut.

To use the table, check down the degrees column until the desired angle is found. Read the numbers opposite for the cut of the roof.

To illustrate:

Assume that a roof is drawn and marked as a 33¾-deg. angle. Check down the list until 33¾ deg. is found. Opposite this read the cut, which is found to be 8 and 12. The cut, therefore, for a 33¾-deg. angle roof is 8 and 12.

TABLE I.—CUTS OF A ROOF IN TERMS OF DEGREES

Degree	Cut	Degree	Cut
2½	½ and 12	46¼	12½ and 12
4½	1 and 12	47¼	13 and 12
7	1½ and 12	48½	13½ and 12
9½	2 and 12	49½	14 and 12
11¾	2½ and 12	50½	14½ and 12
14	3 and 12	51½	15 and 12
16¼	3½ and 12	52¼	15½ and 12
18½	4 and 12	53¼	16 and 12
20½	4½ and 12	54	16½ and 12
22½	5 and 12	54¾	17 and 12
24½	5½ and 12	55½	17½ and 12
26½	6 and 12	56¼	18 and 12
28¼	6½ and 12	57	18½ and 12
30¼	7 and 12	57¾	19 and 12
32	7½ and 12	58½	19½ and 12
33¾	8 and 12	59	20 and 12
35¼	8½ and 12	59¾	20½ and 12
37	9 and 12	60¼	21 and 12
38½	9½ and 12	61	21½ and 12
40	10 and 12	61½	22 and 12
41¼	10½ and 12	62	22½ and 12
42½	11 and 12	62½	23 and 12
43¾	11½ and 12	63	23½ and 12
45	12 and 12	63½	24 and 12

CHAPTER VII

SUMMARY OF ROOF INFORMATION AND BASIC LAYOUT PRINCIPLES

I. General Information:

Four types of roofs
Four kinds of rafters
Four basic rafter cuts
Two applications of term "pitch"

II. Definitions:

Span
Rise
Run
Pitch
Rafters
Ridge
Rafter cuts
Cut of a roof
Shortening
Rafter lengths
Angle of cuts

III. Basic Layout Principles:

Relationship of rafters in plan view
Principle of shortening
Four basic layout problems
Pitch, rise, run, and span relationships
Run of hip, valley, and jack rafters
Principle of similar triangles
Rafter size in relationship to rafter length

IV. Determining Rafter Lengths Mathematically:

Common rafters
Hip and valley rafters

CHAPTER VII

SUMMARY OF ROOF INFORMATION AND BASIC LAYOUT PRINCIPLES

The following pages have been written to serve as a review of the roof principles and rules covered in the previous chapters. No explanation is given for any of the topics covered. The experienced roof framer should be sure he understands fully the detailed description of the various rafter layout problems before he reads this chapter.

This review material is divided into four parts, namely, (1) *general information,* (2) *definitions,* (3) *basic rafter layout principles,* and (4) *determining rafter lengths mathematically.*

GENERAL INFORMATION

1. There are four types of roofs:

    Shed: slant in one direction.
    Gable: slant in two opposite directions.
    Hip: slant in four directions.
    Intersecting: on buildings that have offsets or angles.

2. There are four kinds of rafters:

    Common.
    Hip.
    Valley:
      - Full length.
      - Shortened.

    Jack:
      - Hip.
      - Valley.
      - Hip-valley.
      - Cripple.

3. There are four basic rafter cuts:

   Ridge: at the ridge (a plumb cut).
   Cheek: at top end of hips, valleys, and jacks (an angle cut).
   Seat: at the plate line (a combination of plumb and level cuts).
   Tail: at exposed lower end of rafter (may be any angle but usually a plumb cut or combination of plumb and horizontal cuts).

4. There are three applications of the term "pitch":

   Pitch: mathematical relationship between total rise and span; also mathematical relationship between rise in inches and unit of span (24 in.).
   Equal pitch: roof has the same slant for all its parts.
   Unequal pitch: roof has two different slants or pitches that intersect.

Note: The term "odd pitch" is often applied to total rise and run figures that do not have a common fractional relationship.
To illustrate: 7 ft. 4½ in. rise and 19 ft. 2¼ in. run.

DEFINITIONS

The following definitions are grouped according to meaning rather than alphabetical classification.

1. Span:

   Total span: distance from common rafter seat to opposite common rafter seat.
   Unit of span: always 24 in.

2. Rise:

   Total rise: vertical distance from plate line to ridge.
   Rise in inches: distance roof rises per foot of run.

3. Run:

   Total run: always one-half the total span.
   Unit of run: always 12 in.

4. Rafters:

   Common rafter: the rafter that extends at right angles from plate line to ridge. This rafter is "common" to all types of roofs, hence its name.

Hip rafter: the rafter at the intersection of the four sloping surfaces of a hip roof.

Valley rafter: the rafter that is framed at the intersection of two planes of an intersecting roof.

Supporting valley: a rafter that runs from plate to ridge.

Shortened valley: a rafter that runs from plate line *to* supporting valley.

Jack rafters:

Hip jack: a rafter that is framed against a hip and runs to the plate.

Valley jack: a rafter that is framed against a valley and runs to the ridge.

Hip-valley jack: a rafter that is framed between a hip and valley.

Valley cripple jack: a rafter that is framed between shortened and supporting valley rafters.

5. Ridge: the highest part of the roof (where the ridge cuts of the common rafters meet).
6. Rafter cuts:

Ridge: at the top end of common, hip, valley jack, and full length valley rafters.

Seat: the cut at the intersection of a rafter with the wall plate on which it is to be nailed.

Tail: the cut at the bottom end of any rafter that overhangs to form a cornice.

Cheek:

Single: a slanting plumb cut on jack, valley, and hip rafters to make them fit against the piece to which they are framed.

Double: the same as a single cheek cut except that the slanting cut is in *two* directions.

7. Cut of a roof: This term is used by the carpenter to indicate what figures are required in laying out the common rafter. These figures are always "rise in inches" and 12 in. for an equal pitch roof.
8. Shortening: the amount necessary to shorten each rafter to allow for the thickness of the stock to which it is framed.

9. Rafter lengths:

Mathematical: the exact mathematical length of each rafter with no allowance for shortening.

Note: In the preceding chapters the term "line length" instead of "mathematical length" was used, this term being somewhat easier to understand.

Actual: the actual length of a rafter after allowance has been made for shortening.

10. Angle of the cut:

For common rafters: rise in inches and 12 in.

For hip or valley rafters: rise in inches and 17 in. (when pitch is equal).

For odd-pitch roofs: total rise and total run measurements, reduced to twelfths. (See p. 102.)

BASIC RAFTER LAYOUT PRINCIPLES

1. Relationship of rafters in plan view:

Most of the difficulties of rafter layout encountered by the carpenter apprentice will disappear as soon as he has learned to visualize rafter relationships in their *plan view.* The slope of the roof need not be considered for the moment. The following rules apply to all rafters when considered *in plan view:*

Common rafters intersect the ridge at 90 deg.

Hips, valleys, and jacks of an equal pitch roof intersect the roof member to which they are framed at 45 deg.

Cheek cuts for equal pitch roofs are laid out at right angles to the plumb cut.

Hips, valleys, and jacks for unequal pitch roofs intersect the member to which they are framed *at any angle other than* 45 *deg.* This is the basic reason why unequal pitch roofs are more difficult to lay out than equal pitch roofs.

2. Principle of shortening:

All rafters that are framed against some other roof member must be shortened to allow for the thickness of this member. The rules for shortening are:

All shortening is on the principle of the *plan view;* therefore, all measuring is done at right angles to the plumb cut.

All shortening is on the basis of one-half the thickness of the stock to which the rafter is being framed:

For common rafter, the shortening is one-half the thickness of the stock.

For hips, valleys, and jacks it is one-half the 45-deg. thickness of the stock.

Note: When shortening, a rafter measured at right angles to the plumb cut is automatically decreased in length the correct amount, irrespective of the slope or pitch of the roof.

3. Basic layout information:

Four things must be known about every rafter in order to lay it out:

a. *Mathematical length* (line length).

b. *Angle of the cut* which for the equal pitch roof is rise in inches and 12 in. for the run of the common rafter and rise in inches and 17 in. for the run of the hip or valley rafter.

c. *Shape of the cut,* namely, plumb, level, or cheek cut.

d. *Shortening* (see 2 above).

4. Pitch, rise, run, and span relationship:

One-half the span equals the total run.

The span times the pitch gives the total rise.

5. Run of hip, valley, and jack rafters:

The run of an *equal pitch* roof hip or valley rafter is *always* the diagonal of a square, hence 45 deg.

The run of an *unequal pitch* roof hip or valley rafter is *always* the diagonal of a rectangle and is therefore never 45 deg.

The run of a jack rafter for an equal pitch roof is always equal to its distance from the corner.

6. Principle of similar triangles:

For an equal pitch roof the angle formed by *rise in inches and* 12 *in.* of run is identical with the same angle formed by *total rise* and *total run*. These triangles are known as "similar triangles." Hence if the angle based on 12 in. of run is known, the roof angles for the common rafter are also known.

7. Rafter size in relation to rafter length:

The mathematical or line length of a rafter is *always the same* for any given pitch irrespective of the size rafter stock being used. A rafter cut from a piece of 2 × 10 would be exactly the same length as a rafter cut from a piece of 2 × 4.

DETERMINING RAFTER LENGTHS MATHEMATICALLY[1]

Table II provides a quick method for figuring the lengths of rafters mathematically. The table gives the pitch, cut, and rafter length for a number of different roof pitches. To use the table, proceed as follows:

1. Find the span of the roof by reading the blue print.
2. Divide the span distance by 2 to get the total run.
3. Find the pitch of the roof (usually given on the blue print).
4. Common rafter: Select a constant in Column III for the required pitch, multiply the total run by this constant. The product equals the mathematical length of the common rafter. This should be increased to the next even foot length of lumber.
5. Hip or valley rafter: Select a constant in Column IV and proceed as for the common rafter.

[1] For a detailed explanation of rules to follow when estimating the numerous items required to construct a standard residence see Wilson and Roger's "Simplified Carpentry Estimating," Simmons-Boardman Publishing Corporation, New York.

TABLE II.—LENGTHS OF COMMON, HIP, AND VALLEY RAFTERS PER 12 IN. OF RUN

I Pitch	II Cut	III Length in inches, common	IV Length in inches, hip or valley*
1/12	2 and 12	12 1/8	17 1/8
1/8	3 and 12	12 3/8	17 1/4
1/6	4 and 12	12 5/8	17 3/8
5/24	5 and 12	13	17 3/4
1/4	6 and 12	13 3/8	18
7/24	7 and 12	13 7/8	18 3/8
1/3	8 and 12	14 3/8	18 3/4
3/8	9 and 12	15	19 1/4
5/12	10 and 12	15 5/8	19 3/4
11/24	11 and 12	16 1/4	20 1/4
1/2	12 and 12	17	20 3/4
13/24	13 and 12	17 3/4	21 3/8
7/12	14 and 12	18 1/2	22
5/8	15 and 12	19 1/4	22 5/8
2/3	16 and 12	20	23 3/8
17/24	17 and 12	20 3/4	24
3/4	18 and 12	21 5/8	24 3/4
19/24	19 and 12	22 1/2	25 1/2
5/6	20 and 12	23 3/8	26 1/4
7/8	21 and 12	24 1/8	27
11/12	22 and 12	25	27 3/4
23/24	23 and 12	26	28 5/8
Full	24 and 12	26 7/8	29 3/8

* These figures are the length of the hip or valley rafter for 12 in. of common rafter run.

CHAPTER VIII

THE STEEL SQUARE AND ITS USE

I. The Carpenter's Steel Square
II. Laying Out Roof, Stair, and Polygon Cuts
III. Determining Rafter Lengths by Measuring Diagonal of Cut:

Common rafter
Hip or valley rafter

IV. Determining Rafter Lengths by the "Step-off" Method:

Common rafter
Hip or valley rafter

V. Applying the Principle of the Similar Triangle:

Roofs
Stairs
Polygons

VI. Using the Tables on the Steel Square:

Linear measurements
Hundredths scale
Octagon scale
Board measure
Brace lengths
Rafter lengths:
  Common
  Hip or valley
  Jack
Rafter cheek cuts:
  Hip or valley
  Jack
Sheathing cuts
Polygon cuts

CHAPTER VIII

THE STEEL SQUARE AND ITS USE

Roof framing and stair building layout problems are primarily based on geometric principles. The steel square greatly simplifies this layout work. It can also be used to solve other building construction problems. The following pages briefly describe how to use the square in the solution of these problems and include a number of roof framing short cuts not given in the previous chapters. For the experienced carpenter the material can be considered as a quick reference on most of the ordinary roof and stair layout situations. *It must be emphasized that the carpenter should familiarize himself with the basic principles as described in previous chapters before using the short-cut methods.*

THE CARPENTER'S STEEL SQUARE[1]

The carpenter's steel square is a hand tool constructed of steel with a tongue that is 1½ in. wide and varies in length (usually 16 in. long). The blade is always 2 in. wide and 24 in. long. For convenience in handling it is usually held with the

[1] The manufacture of steel carpenter's squares in the United States was begun by Silas Hawes, a blacksmith, at South Shaftsbury, Vermont, in 1817. This manufacturing business was later the foundation of the Eagle Square Company which was incorporated in 1876. The first squares were manufactured from discarded saw blades and were later drawn under a trip hammer from steel bar stock when that became available.

In 1902 the Nicholls Manufacturing Company, at Ottumwa, Iowa, began the practice of stamping the framing rules on the square.

It is interesting to note that as late as 1910 these steel squares were hand welded, while at present they are blanked out of sheets in one piece.

In 1916 the Eagle Square Company was acquired by the Stanley Rule Level Company, which is today known as Stanley Tools, with headquarters at New Britain, Connecticut.

tongue in the left hand and the outside corner of the tool away from the carpenter (see Fig. 12).

Examining the square, one finds that it is not stamped alike on both sides. In fact the markings vary with the different makes of squares. Generally, however, on the face of the tool (maker's name is stamped on the face side) the inches are subdivided into quarters, eighths, and sixteenths, while the back of the square is marked into twelfths.

For some of the more difficult problems in roof framing and stair building the $\frac{1}{12}$ scale is very convenient and is the one most commonly used.

Many jobs are laid off by means of the square that are larger than the tool itself. By using the $\frac{1}{12}$ scale it is possible to read directly in feet and inches. One inch equals 1 ft. and $\frac{1}{12}$ in. equals 1 in.

If the object to be framed is too large for the scale, reduce to $\frac{1}{24}$ scale ($\frac{1}{2}$ of $\frac{1}{12}$), proceed as before, and double the answer.

Note: Holding the square in the position as suggested above will bring the twelfth scale up where it can be seen by the carpenter.

For convenience the problems in this chapter are divided as follows:

Laying out roof, stair, and polygon cuts.
Determining rafter lengths by measuring diagonal of cut.
Determining rafter lengths by the "step-off" method.
Applying the principle of the similar triangle.
Using the tables on the steel square.

LAYING OUT ROOF, STAIR, AND POLYGON CUTS

A. For the following cuts let one number on the tongue of the square represent the roof *rise*, and let the other number on the body, or blade, represent the *run* of the roof.

1. *Laying out rafter cuts:*
   a. To lay out a *plumb* cut of a common rafter: use rise and run (12 in.); cut on rise.
   b. For the *plumb* cut of a hip or valley rafter on an equal pitch roof: use rise and run (17 in.); cut on rise.

c. To lay out the *plumb* cut of an unequal pitch valley rafter: use total rise and total run; cut on total rise.

d. To lay out the *level* cut of a common rafter: use rise and run (12 in.); cut on run.

e. For the *level* cut of hip or valley rafter on an equal pitch roof: use rise and run (17 in.); cut on run.

f. To lay out the *level* cut of an unequal pitch roof valley rafter: use total rise and total run; cut on total run.

g. For the tail cut of a hip rafter which is to match a *square* tail cut on a common rafter: use *one-half* the rise of the common rafter and 12; cut on rise. Be sure to work from the top edge of the rafter. This will make the long part of the cut at the top edge.

h. To lay out a *cheek* cut for a jack rafter: use run (12 in.) and length; cut on length.

Note: The word "length," when applied to a jack rafter cheek cut means the *length of rafter per foot of run* (the diagonal of the cut). On the job the term "*run and length; cut on length*" is often used by the carpenter to indicate figures required to lay out a cheek cut. The figures used for "length" are different for each roof pitch (see Table II, p. 125).

When laying out the jack rafter cheek cut, the "diagonal of cut" measurement is easily obtained by measuring across the square from the rise and run figures of the common rafter (see Fig. 7). Some steel squares show the actual figures to use (see p. 148).

i. To lay out a hip or valley rafter *cheek* cut (equal pitch roof): use run (17 in.) and length; cut on length.

Note: The term "length" in this case means the length of the hip rafter for each 17 in. of hip rafter run. The run figure will extend beyond the limits of the 16-in. tongue of the square; hence both run and length figures must be reduced one-half.

To illustrate:

On an 8 and 12 cut the length of the hip rafter for each 17 in. of hip rafter is 18¾ in. The side cut for the hip or valley cheek cut would be 18¾ and 17, which are reduced to 9⅜ and 8½ in order to use the square.

j. To lay out a jack rafter *cheek* cut (unequal pitch roof): Minor span rafters: use one-half the *major* span (see Fig. 81) and length of common rafter of the *minor* span; cut on length.
Major span rafters: use one-half the *minor* span and length of common rafter of the *major* span; cut on length.

2. *Laying out gable stud cuts:*
To lay out a cut for the top end of a gable stud: use rise and run (12 in.); cut on rise.
3. *Laying out collar tie cuts:*
To lay out the end of a collar tie to make it conform to the slope of the roof: use rise and run (12 in.); cut on run.

B. For the following cuts use run *or* rise and length (diagonal of the cut).

1. *Laying out sheathing cuts:*
    a. To lay out the miter or top cut of a sheathing board to make it fit on top of a hip or valley rafter: use run (12 in.) and length; cut on run.
    b. To lay out the face or edge cut of a sheathing board: use rise and length; cut on rise.
2. *Laying out purline cuts:*
    a. To lay out the miter or top cut of a purline: use run (12 in.) and length (diagonal of cut); cut on run.
    b. To lay out the face or side cut of a purline: use rise and length (diagonal of cut); cut on rise.

C. For the following cuts use tread rise and tread run:

1. *Laying out stair cuts:*[1]
    a. To lay out a riser or plumb cut: use tread rise and tread run; cut on rise.
    b. To lay out a tread or level cut: use tread rise and tread run; cut on run.

[1]For a detailed description of stair layout see "Simplified Stair Layout," by Wilson and Werner. Delmar Publishers, Inc., Albany, N.Y.

c. To lay out the top and bottom cuts on a stair horse or stringer: use tread rise and tread run; mark on rise for plumb cut and run for level cut.

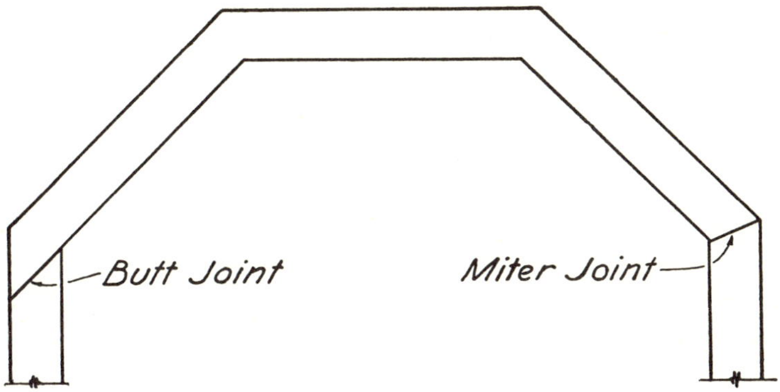

Fig. 116.

D. For the following polygon miter and butt cuts use figures from Table III selected according to the kind of polygon. The miter and butt cuts (see Fig. 116) for various polygons are as follows. Always cut on the first-named figure.

TABLE III.—POLYGON CUTS

Number of sides	Name of polygon	Miter joint cut	Butt joint cut
3	Triangle	20¾ and 12	6¹⁵⁄₁₆ and 12
4	Square	12 and 12	Use right angle of square
5	Pentagon	8¾ and 12	3⅞ and 12
6	Hexagon	6¹⁵⁄₁₆ and 12	6¹⁵⁄₁₆ and 12
7	Heptagon	5¾ and 12	9⁹⁄₁₆ and 12
8	Octagon	5 and 12	12 and 12
9	Nonagon	4⅜ and 12	14⁵⁄₁₆ and 12
10	Decagon	3¹⁵⁄₁₆ and 12	16½ and 12
11	Undecagon	3½ and 12	18¹¹⁄₁₆ and 12
12	Dodecagon	3¼ and 12	20¾ and 12

Note: Some steel squares give miter cuts for polygon figures that are not the same as above. Actually there is no difference. The above cuts are all based on 12 in. instead of some other number.

DETERMINING RAFTER LENGTHS BY MEASURING DIAGONAL OF CUT

Rafter lengths can easily be determined on the square by letting the tongue of the square represent one of the measurements of the rafter angle and the blade the other dimension and measuring diagonally from one figure to the other. This measuring is simplified if a pair of square gages is used to locate each dimension accurately, and a second square is used to measure from gage to gage (see Fig. 7).

Various rafter lengths are obtained as follows:

1. To Find the Length of a Common Rafter.—(Using 1/12 Scale on Steel Square.) Use one-half the span measurement on one part of the square and the total rise measurement on the other part. Measure the diagonal distance between these points. This is the 1/12 scale length of the common rafter. Each inch represents a foot and each twelfth represents an inch.

2. To Find Length of an Equal Pitch Roof Hip or Valley Rafter.—Use one-half the span measurement and the common rafter length measurement and proceed as outlined above for the common rafter. Result equals the 1/12 scale length of the hip or valley rafter.

3. To Find the Length of an Unequal Pitch Roof Valley Rafter. *a. Using Major Span Measurements.*—(Figure 81.) Use one-half the *major span* distance and the length of the common rafter of the *minor* span and proceed as outlined above. Result equals scale length of the valley rafter.

b. Using Minor Span Measurements.—(Figure 81.) Proceed as outlined above for the major span but use the reverse measurements, namely, one-half the *minor* span measurement and the length of the common rafter of the *major* span.

Note: The valley rafter length is the same by either method *a* or *b*. To avoid errors it is advisable to check the length found by one method against the length obtained by the other.

DETERMINING RAFTER LENGTHS BY THE "STEP-OFF" METHOD

Probably the most commonly used method in laying off rafters is that of "stepping off" the required distance with the steel square using the $\frac{1}{12}$ scale. This is an easy method when the unit of run is a standard length. For the common rafter this is always 12 in. For hips and valleys the unit of run for an equal pitch roof is always 17 in. Therefore the number of "steps" taken with the square when laying off either common, valley, or hip rafters is the same *as the number of feet in the total run of the common rafter.*

To illustrate:

If a common rafter had a run of 16 ft. there would be 16 steps required for the common rafter; likewise 16 steps would be used for the hip or valley rafter. The hip or valley run, however, is 17 in.; that is, it has 5 in. more length in each step.

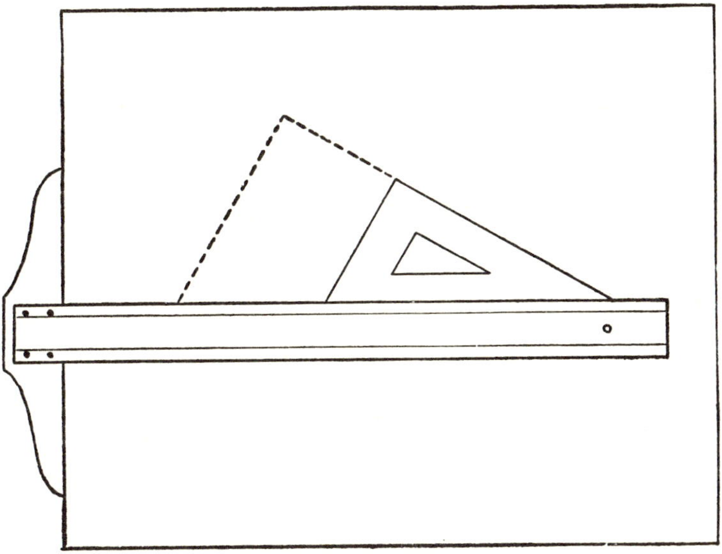

Fig. 117.

1. Laying Off Common Rafters by the "Step-off" Method.—(Figure 14.) Set the square gages to the cut of the roof and move the square as many times as there are feet in the total run. A part of a foot of run is laid off by selecting on the body of the square the exact number of inches of run required.

2. Laying Off Hip or Valley Rafters by the "Step-off" Method.—(Figure 29.) Set the square gages to rise in inches and 17 in.; then move the square as many times as there are feet in the total run of the *common* rafter. A part of a foot of run is laid off by using the fractional relationship that the part is to 12 in. and applying it to the 17 in.

To illustrate:

Six inches of run is ½ ft. Therefore ½ of 17 or 8½ in. would be laid off on the body of the square. Likewise 4 in. (⅓ of 12 in.) of common rafter run equals ⅓ of 17 in. or 5 11/16 in. of hip or valley run. The quickest way to find the fractional part of 17 in. is to measure across the square from one number on the tongue to the same number on the blade. The number selected is the fractional part of the common rafter run.

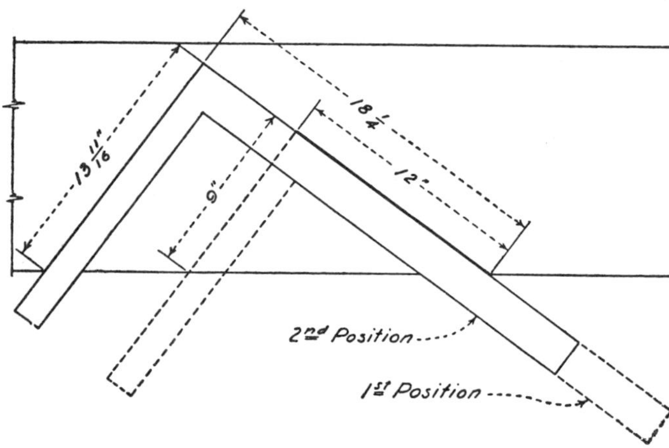

Fig. 118.

Note: If total rise and total run measurements are employed when laying off a rafter, the 1/12 scale is used, and the square is *always* stepped twelve times irrespective of the pitch of the roof or the amount of run. This is necessary because each step measures exactly one-twelfth of the total rafter length.

APPLYING THE PRINCIPLE OF THE SIMILAR TRIANGLE

Many problems involving angles can be solved by using the geometric principles underlying the similar triangle. Similar triangles have identical angles but are not of the same size. A small triangle can be "projected" into a larger one or vice versa with all angles of both remaining the same.

This projection as developed by the draftsman is illustrated in Fig. 117. The carpenter, however, can accomplish the same result with the steel square. This procedure is known as "*slipping the square*" and can be used to good advantage on a number of construction problems, as it not only shortens the time required to do the work but is a practical method to use on the job.

A smooth wide board is preferable for the following layout problems, and for accuracy a knife should be used when marking the various angles.

How to Slip the Square.—(Figure 118.) 1. Selecting figures to represent a right-angle triangle lay out this angle, using the $\frac{1}{12}$ scale on the square.

2. Determine one dimension of the larger right triangle.

3. Letting one edge of the square exactly coincide with one of the lines of the smaller triangle, "slip" the square so that it represents one dimension of the larger triangle.

4. Read the figure on the other leg of the square. This will be the second dimension of the new right-angle triangle.

To illustrate:

Assume a right-angle triangle of 9 in. and 12 in. The run (base) of the new angle is $18\frac{1}{4}$ in. What is the rise (altitude) for this run which will have the same mathematical relationship as 9 in. and 12 in.?

Solution: Lay out the 9 in. and 12 in. angle with the steel square and mark on the run (12) side. "Slip" the square to $18\frac{1}{4}$ in. The new rise is $13\frac{11}{16}$ in. An angle of $13\frac{11}{16}$ and $18\frac{1}{4}$ in. is identical to the angle formed by 9 in. and 12 in.

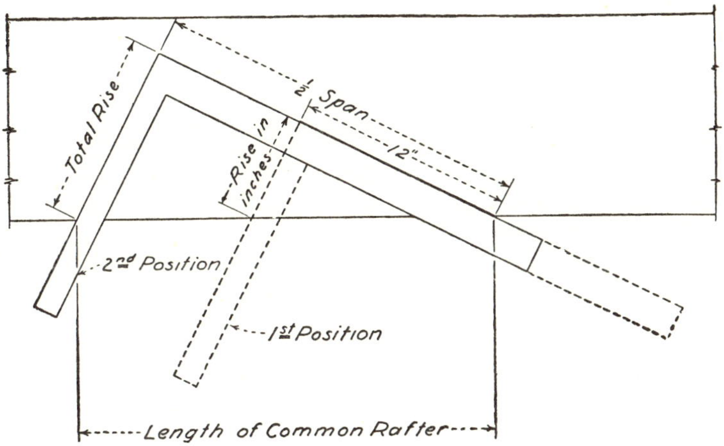

FIG. 119.

Construction Problems Solved by "Slipping the Square." 1. *Length of Common Rafter.*—(Figure 119.) Set square to rise and run and mark on run. Slip the square to one-half span distance. Read total rise. Measure from new rise figure on the tongue of square to one-half span figure on the body of the square. This is the $\frac{1}{12}$ scale length of the common rafter.

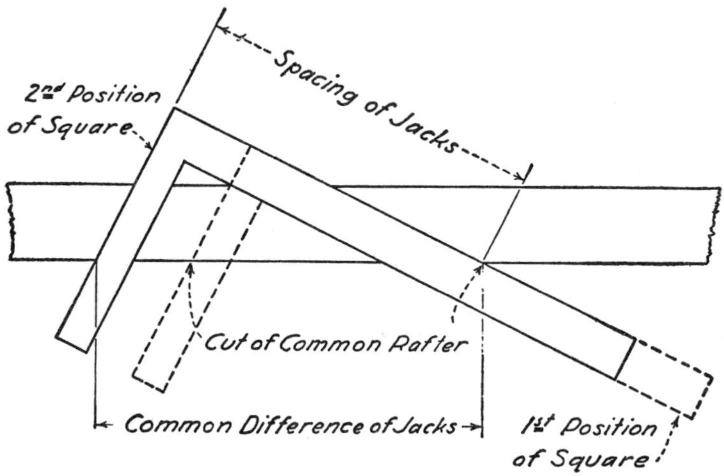

FIG. 120.

2. *Total Rise of Roof.*—(Figure 119.) Set square to rise and run and mark on run. Slip square to one-half span. Read total rise.

3. *Common Difference in Length of Jack Rafters.*—(Figure 120.) Set square to rise and run (12 in.) and mark on run. Slip square to figure representing the spacing of the jacks and measure diagonal of new angle. This is the common difference measurement.

4. *Common Difference in the Lengths of Gable Studs.*—(Figure 121.) Set square to rise and run (12 in.) and mark on run. Slip square to spacing of studs and read other dimension on the tongue of the square. This is the common difference in lengths of gable studs.

5. *Length of Collar Tie.*—(Figure 122.) Set square to rise and run (12 in.) and mark on *rise.* Slip square to figure

representing vertical distance collar tie is down from ridge (as shown on the blue print). Read new run which is *one-half* length of collar tie. Double result to obtain full length. This length can be either the top or lower edge of the tie. For the angle cut at the end of the tie use the cut of the roof and cut on run.

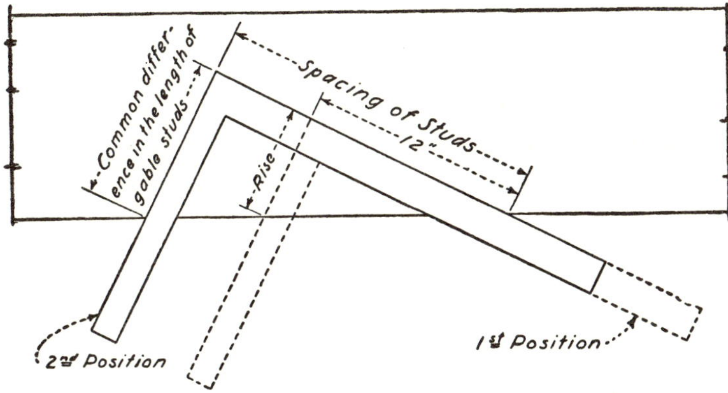

FIG. 121.

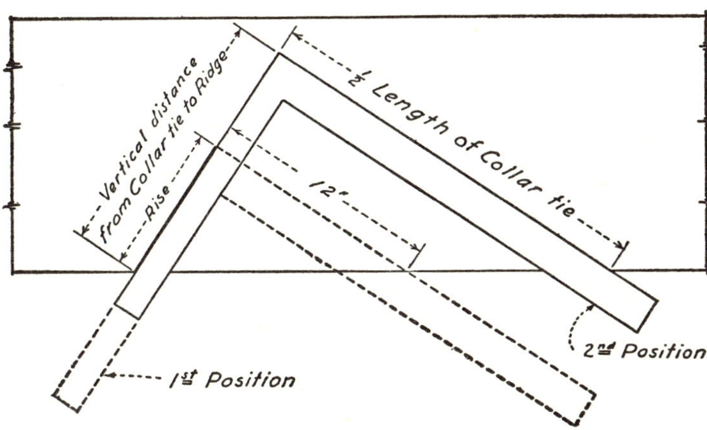

FIG. 122.

6. *Cut of an Odd or Unequal Pitch Roof in Terms of* 12 *in. of Run.*—(Figure 123.) Set square to total rise and total run figures of the odd or unequal pitch roof, mark on run. Slip square to 12 in. and read new rise. This number and 12 is the cut of the roof.

7. *Distance to Raise Plate on Minor Span of an Unequal Pitch Roof.*—(Figure 124.) Several steps are required as follows (read pp. 89, 90):

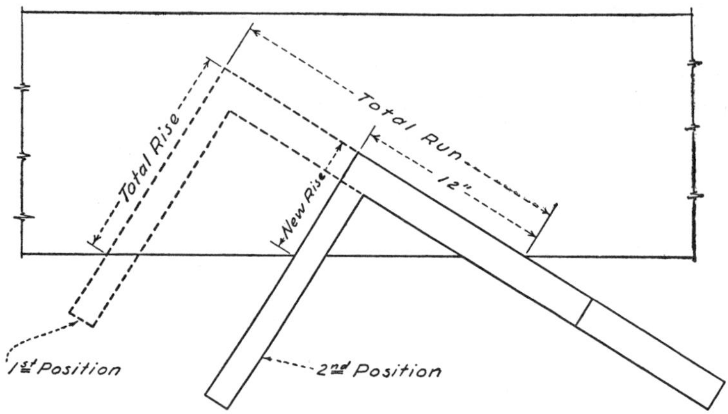

FIG. 123.

Find amount of cornice "drop" of the main roof (major span) by setting square to cut of roof and then slipping it to the run of the cornice; new rise is also the drop of the cornice of the major span.

Add *drop of the cornice* to total rise of the main roof and *run of cornice* of the main roof to the run of the minor span. *Increased* total rise and *increased* total run is the cut of the common rafter of the minor span.

Using these figures as the angle, slip the square to the run of the cornice; read new rise figure, which is the drop of the cornice of the minor span.

Subtract drop of the cornice of the main roof from the drop of the cornice of the minor roof. Result is amount to raise the plate of the minor roof.

To illustrate:

Assume a building layout as shown in *A*, Fig. 124. Major span is 38 ft. with a ¼ pitch. Minor span is 20 ft. Cornice run is 16 in. How much should the plate of the minor roof be raised?

Solution: The major rise is ¼ of 38 ft. or 9 ft. 6 in. The cut for a ¼ pitch is 6 and 12. Set the square to 6 and 12; slip it to 16 in. (the cornice run) and read the new rise, which is also the cornice drop or 8 in. (see Fig. 124*B*).

The rise cut of the minor roof is 9 ft. 6 in. (major roof rise) *plus* the cornice drop of 8 in. or 10 ft. 2 in. The *run* is one-half the minor span or 10 ft. 0 in. *plus* the cornice run of 16 in. or 11 ft. 4 in.

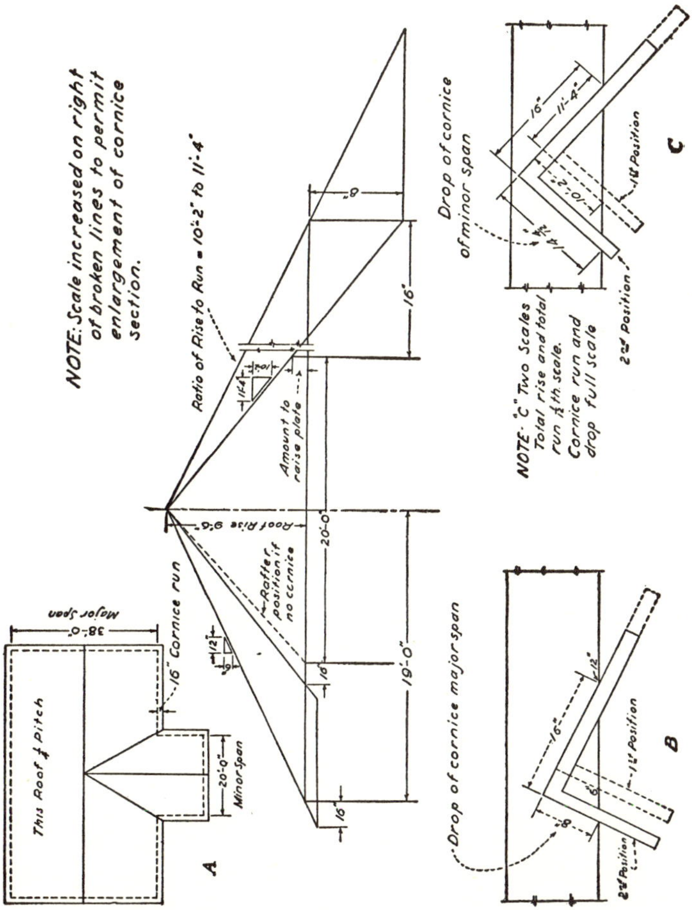

FIG. 124.

Set the square gages to 11 ft. 4 in. (increased total run) and 10 ft. 2 in. (increased total rise) and mark on the run side. "Slip" the square to the cornice run and read the new rise, which is 14⁴⁄₁₂, called 14⅜ in. This is actually the drop of the cornice of the minor roof (see Fig. 124*C*). The amount to raise the plate of the minor roof is 14⅜ in. less 8 in., which equals 6⅜ in.

8. *Tread Rise of a Stairway.*—(Figure 125.) Set square to total rise on the tongue and number-of-risers on blade and

mark on the number-of-risers edge of the square. Slip square to 12 in. Read number on tongue of square. This is the tread rise and the amount to set the dividers when laying off the stair rod.

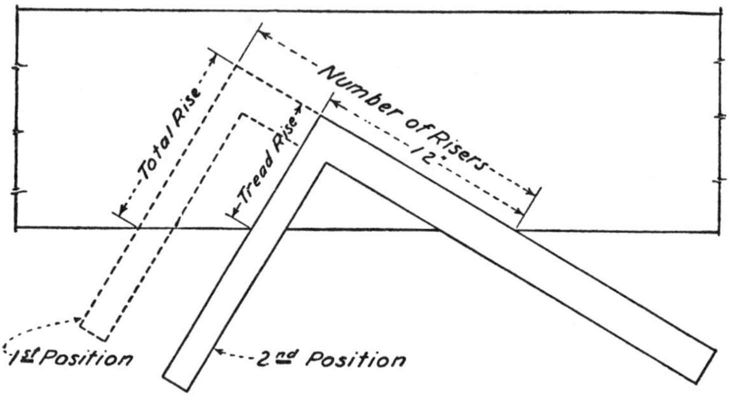

FIG. 125.

USING THE TABLES ON THE STEEL SQUARE

Most of the steel squares made by reliable manufacturers have a number of different tables stamped on the surface of the tool to provide the carpenter with a quick method for determining what figures to use for various angle cuts, rafter lengths, and roof cuts.

The tables on each make of square may vary in form, and some squares have tables not found on others. If two makes of squares are compared it may also be found that different numbers have been given for the same angles. This means that a different *run* figure has been selected for a given angle, which likewise results in a different *rise* figure. However, if two angles are laid out, using a different square for each, and then compared, they will be found to be identical in shape.

There is no standard practice followed when stamping figures on the square. The tables on the square must be carefully studied before any layout work is done to be sure which figure to use. It must again be emphasized that the tables on a steel square should be used *only* when the carpenter understands the principles of roof layout.

All figures given for jack rafter cheek cuts are based on the rule of "run (12 in.) and length (diagonal of the cut); cut on length." Sheathing face cuts are exactly the reverse, namely, "run and length; cut on *run*." Hip and valley rafter cheek cuts are based on the unit of run for these rafters, which is 17 in. and the length of the rafter per each 17 in. of run. Again the rule: run and length; cut on length but with a set of figures pertaining to the hip or valley rafter.

In the following pages the tables on several different makes of squares are described under the same general heading. These tables can be classified as follows:

1. Linear measurement scale.
2. Octagon scale.
3. Board measure.
4. Brace lengths.
5. Common rafter length per foot of run.
6. Hip or valley rafter length per foot of run.
7. Actual length of common rafters.
8. Actual length of hip or valley rafters.
9. Common difference in the length of jack rafters.
10. Side (or cheek) cut of jack rafters.
11. Side (or cheek) cut of hip or valley rafters.
12. Sheathing cuts.
13. Polygon cuts.

Each of the above tables is described in detail in the following paragraphs:

1. Linear Measurement Scales.—(Figure 126.) *a. Linear Measurements.*—The figures given on the inside and outside edges of the various squares are used for measuring linear inches and feet. The difference in the several scales given is in terms of the inch divisions. Some are divided into spaces as small as 1/32 in., some into eighths, and on one edge of all squares made primarily for carpenters the inches are divided into twelfths (1/12). The 1/12 scale is the one most used by carpenters when laying out rafters and stairs. On one or two squares, usually on the tongue, the inches are divided into tenths (1/10). This table is valuable when converting a decimal part of an inch to a fraction.

Note: Any carpenter who expects to become a good roof framer should possess a square that has a 1/12 scale.

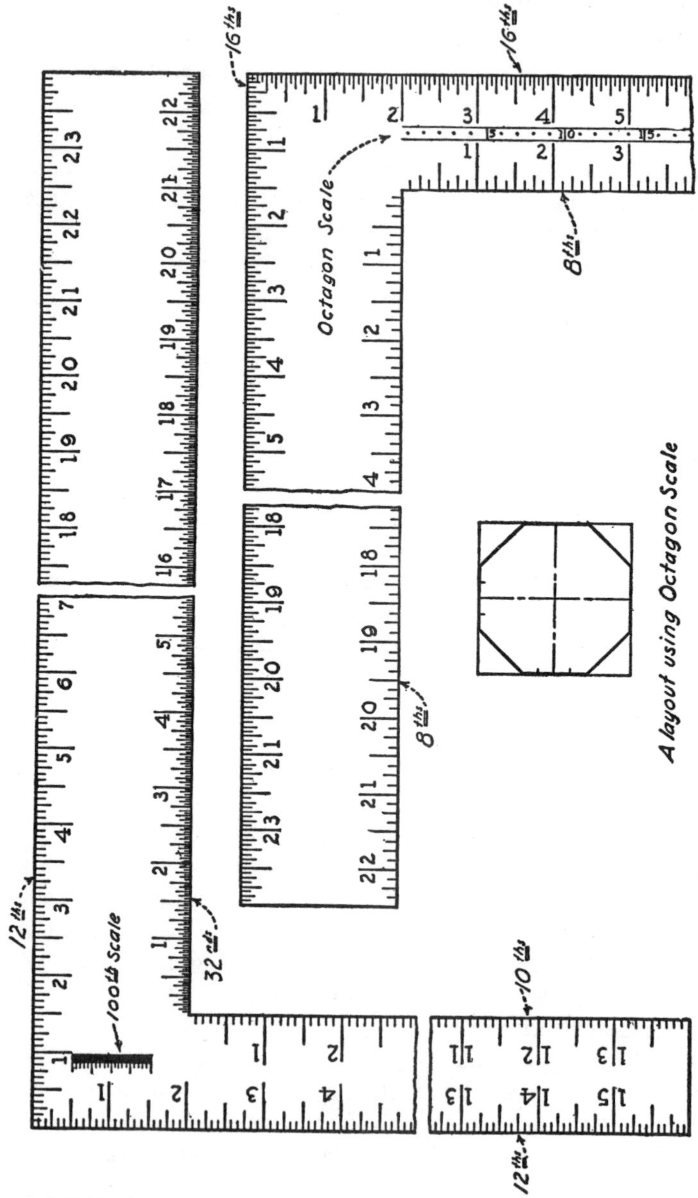

A layout using Octagon Scale

Fig. 126.

b. The Hundredth Scale.—(Figure 126.) To enable the carpenter to convert a decimal part of an inch to a fraction, the 1/100 scale is stamped on the back and near the heel or corner of each of the Nicholls or Stanley squares.[1] To use the scale, set a pair of carpenter's dividers on it to the required decimal, transfer measurements to one of the standard inch scales on the square and read the fractional answer.

2. The Octagon Scale.—(Figure 126.) The octagon scale, if on the square, is found on the face and in the center of the tongue. Note that each major division is in multiples of 5 and that each one is again divided into 5 parts. The purpose of this scale is to provide a quick method of laying out an octagon cut on the end of a timber such as a 10 × 10 in. or 14 × 14 in., etc.

To use the scale proceed as follows:

1. Select the timber to be laid off and measure its dimensions.
2. Mark a center line each way on one end of the piece of lumber, forming a cross.
3. Set a pair of carpenter's dividers to as many parts on the octagon scale as the size of the timber. If the timber is 4 × 4 in., then set to 4 parts; if 12 × 12 in., set to 12 parts, etc.
4. Lay off this distance (as set on the dividers) *each side* of the ends of both center lines. This will have to be done four times as there are four line ends.
5. Connect these points by a diagonal or 45-deg. line, and the layout work is complete.

3. Board Measure Scale.[2]—(Figure 127.) The board measure scale is stamped on the body of the Stanley square and is used to determine the number of board feet of lumber for various width boards and for lengths from 2 to 24 ft.

The key to this table is found under the figure 12, which is on the outside edge of the square blade. Each number beneath

[1] The Nicholls square is manufactured by the Nicholls Manufacturing Company, Ottumwa, Iowa. The Stanley square is manufactured by Stanley Tools, New Britain, Connecticut.

[2] For a detailed description on how to figure board feet of lumber see Chap. VII in Wilson and Rogers, "Carpentry Mathematics," McGraw-Hill Book Company, Inc., New York.

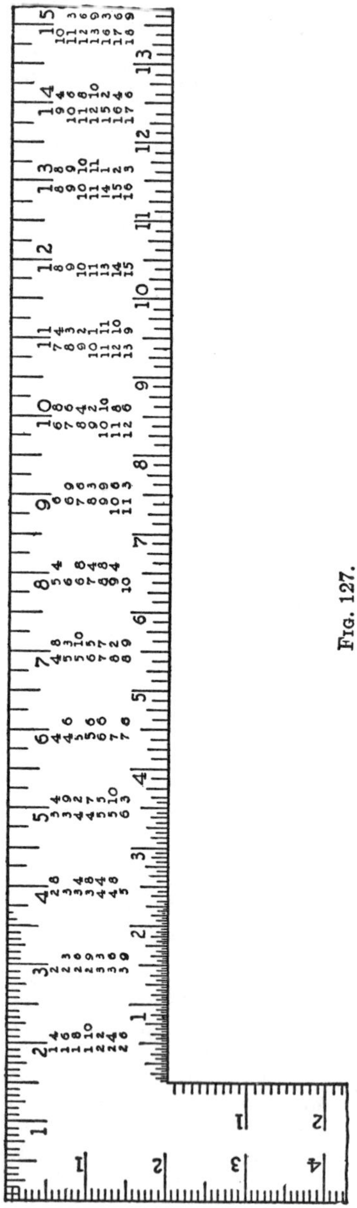

FIG. 127.

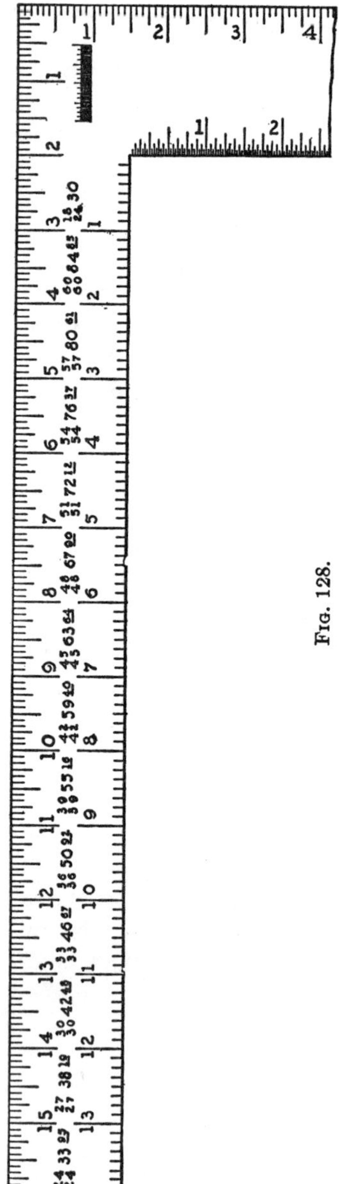

FIG. 128.

12 represents the *width* of a board. Each number on the edge of the blade from 2 to 24 represents the *length* of a board.

To use the table or scale proceed as follows:

1. Select (under the figure 12) the number representing the width of the board to be converted into board feet.

2. Follow horizontally along this set of figures to the figure under the number representing the length of the board.

3. Read this number, which represents the total board feet in the piece of lumber having the selected dimensions. The board foot answer is written 9 4, 11 8, or 12 10, which means $9\frac{4}{12}$ or $11\frac{8}{12}$ or $12\frac{10}{12}$ ft.

To illustrate:

How many board feet are there in a piece of lumber 1 in. × 8 in. × 14 ft.?

Solution: Reading under the number 12 it will be seen that the figure 8 (the width of the board) is in the first line. Follow this line to the figure under the number 14 and read the answer, 9 4 or $9\frac{1}{3}$ feet.

Note: The thickness of the board is always considered to be 1 in. If the piece to be figured is $1\frac{1}{4}$ in. thick, add one-fourth to the answer given on the square; if 2 in., double the answer, etc.

4. Forty-five-degree Brace Lengths.—(Figure 128.) The brace length table is found on the tongue of the Stanley square. A typical example of the different size angles stamped on the square is as follows: $\frac{36}{36}$ 50.92. This means that if the ends of a 45-deg. brace are placed 36 in. each way from the inside corner of the angle, the brace will be 50.92 in. or $50\frac{7}{8}$ in. long. An angle $\frac{54}{54}$ will require a brace 76.37 or $76\frac{3}{8}$ in. long.

To use the table proceed as follows:

1. Determine the distance the ends of the brace are to be placed from the corner.

2. Select the set of figures that corresponds to the distance and read the length of the brace.

3. Convert the decimal to a fraction by one of these methods:

a. Using the $\frac{1}{100}$ scale as given on page 143.
b. Multiplying the decimal by 16 (to convert to 16ths).
c. Using the decimal equivalent Table IV on page 146.

Note: The 45-deg. cut at the end of the brace can be laid out with the square by using the same numbers on the tongue and body, such as 10-10, 14-14, 8-8, etc.

The following table gives equivalent fractions for many of the decimals found on the standard square:

TABLE IV.—DECIMAL EQUIVALENTS

Decimal	Equivalent fraction	Decimal	Equivalent fraction
0.0156	1/64	0.5156	33/64
0.0312	1/32	0.5312	17/32
0.0468	3/64	0.5468	35/64
0.0625	1/16	0.5625	9/16
0.0781	5/64	0.5781	37/64
0.0937	3/32	0.5937	19/32
0.1093	7/64	0.6093	39/64
0.1250	1/8	0.6250	5/8
0.1406	9/64	0.6406	41/64
0.1562	5/32	0.6562	21/32
0.1718	11/64	0.6718	43/64
0.1875	3/16	0.6875	11/16
0.2031	13/64	0.7031	45/64
0.2187	7/32	0.7187	23/32
0.2343	15/64	0.7343	47/64
0.2500	1/4	0.7500	3/4
0.2656	17/64	0.7656	49/64
0.2812	9/32	0.7812	25/32
0.2968	19/64	0.7968	51/64
0.3125	5/16	0.8125	13/16
0.3281	21/64	0.8281	53/64
0.3437	11/32	0.8437	27/32
0.3593	23/64	0.8593	55/64
0.3750	3/8	0.8750	7/8
0.3906	25/64	0.8906	57/64
0.4062	13/32	0.9062	29/32
0.4218	27/64	0.9218	59/64
0.4375	7/16	0.9375	15/16
0.4531	29/64	0.9531	61/64
0.4687	15/32	0.9687	31/32
0.4843	31/32	0.9843	63/64
0.5000	1/2	1.0000	1

To use the table proceed as follows:

1. Read the decimal answer on the steel square and locate the nearest decimal on the table.

2. Read the equivalent fraction for this decimal.

5. Common Rafter Length per Foot of Run.—(Figures 129 and 130.) The length of the common rafter per foot of run is the diagonal of the cut of the roof. The table giving this information is on the body of both the Nicholls and Stanley squares and is labeled the same as the heading above.

Each number from 2 to 18 in. on the edge of the body of the square is considered to be *the roof rise in inches;* the 12 in. of run (always standard) is assumed. *The figure below the rise number is the length of the common rafter per each* 12 *in. of run.*

To illustrate:

What is the length of the rafter per foot of run for a roof that has a 7 and 12 cut?

Solution: Reading horizontally in the proper space that lines up with the heading at the end of the table and selecting the figure under the number 7, the answer is 13.89 (the diagonal of 7 and 12). If the cut of the roof is 9 and 12 the length per foot of run will be 15 in.

Note: If the answer is a decimal, convert to inches by using Table IV on page 146 or use a pair of dividers on the 1⁄100 scale as described above. An even quicker method is to multiply the decimal by 16 to convert it to sixteenths, which is sufficiently accurate for all practical purposes.

To illustrate:

What is the fractional equivalent of 14.42 (the diagonal of 8 and 12)?

Solution: 0.42 × 16 equals 6.72. Using the first number only, that is, 6, call it 6⁄16 or 3⁄8. Therefore 14.42 equals 14 3⁄8. If a more accurate answer is needed, convert to thirty-seconds by multiplying by 32.

To use the table on the square proceed as follows:

1. Determine the cut of the roof.

2. Select the rise figure on the edge of the square.

3. Read the length of the common rafter per foot of run given under this rise number. Be sure to select the correct horizontal space.

Note: Multiplying the length of a rafter per foot of run by the total run of the common rafter will give the mathematical length of the rafter.

To illustrate:

For a 5 and 12 roof, if the common rafter run is 11 ft. 0 in. the length

FIG. 129.

FIG. 130.

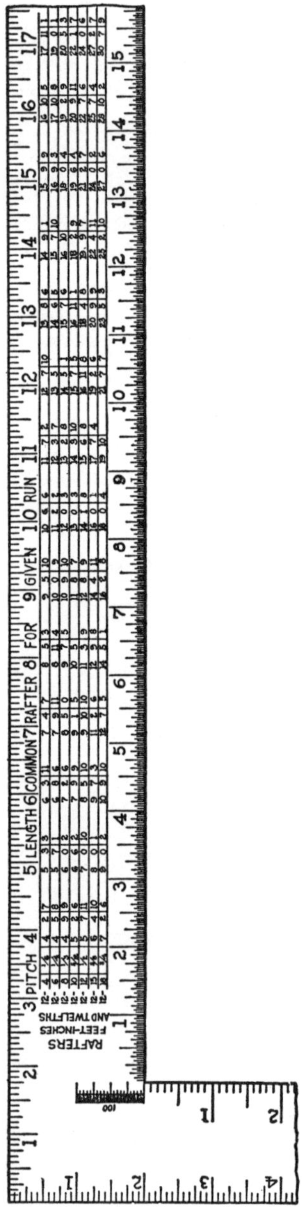

FIG. 131.

of the rafter will be 11 × 13(143 in.) or 11 ft. 10 in. (13 is diagonal of 5 and 12).

6. Hip or Valley Rafter Length per Foot of Run.—(Figures 129 and 130.) The length of the hip or valley rafter per foot of run is found in exactly the same manner as described for the common rafter. This table is located on the square directly below the common rafter per foot of run figures.

To use the table proceed as follows:

Follow all instructions as given for the common rafter but read the hip or valley rafter figures.

7. Actual Length of Common Rafters.—(Figure 131.) The Sargent steel square[1] gives the length of a common rafter for various pitches and for run measurements within the limits of the square. The reading of this table is basically the same as for other tables. Proceed as follows:

1. On the edge of the square select a figure that represents the *run* of the common rafter.
2. At the end of the table select the correct pitch.
3. Under the run figure and in the space that represents the pitch will be found the length of the common rafter.

To illustrate:

What is the length of the common rafter for a 5/8 pitch roof if the building is 30 ft. wide?

Solution: If the span is 30 ft. the common rafter run is 15 ft. Under the figure 15 and in the 5/8 pitch space will be found the answer 24 0 2. This means 24 ft. 0 2/12 in. (The first figure is *feet,* the second is *inches,* and the third is *twelfths of an inch.*)

8. Actual Length of Hip or Valley Rafters.—(Figure 132.) On the back of the same square (Sargent) is a table for hip or valley rafters similar to the one for common rafters as described above. To use this table proceed as follows:

1. Duplicate all steps taken for the common rafter but use the hip or valley table.

9. Common Difference in the Length of Jack Rafters.[2]—(Figures 129 and 130.) Each jack rafter on an equal pitch

[1] This square is manufactured by Sargent & Company, New Haven, Connecticut.

[2] See page 42 for detailed description on how to use the common difference figure when laying out jack rafters.

roof, when the rafters are equally spaced, is always an exact mathematical length longer or shorter than the jack rafter next to it. The amount of common difference is based on the spacing of the jacks and the pitch of the roof.

Note: Jack rafters are usually spaced either 16 or 24 in. on centers. The 16-in. spacing is used if there is plastering on the underside of the rafters or if the roof load, such as tile, requires closer rafter spacing than the standard 24 in. In both instances the 16- and the 24-in. spacing make it possible for some of the rafters to be placed against the ceiling joists enabling the carpenter to nail these two members together, thereby increasing the strength of the building. The tables on the Nicholls and Stanley steel squares cover both 16- and 24-in. spacings.

The method of reading each table is the same. The cut of the roof is selected from the standard inch linear measurement table on the edge of the square, each figure representing the *rise* of the roof, the 12 in. of run always being assumed. Under each rise will be found the common difference in the length of each jack rafter for that pitch roof.

To illustrate:

If a roof has a cut of 10 and 12, what is the common difference in the length of the jack rafters if they are spaced 16 in. on centers?

Solution: Under the figure 10 in the correct space will be found the figure 20.83 (see Fig. 129) or 20⅞ in. (see Fig. 130), which is the common difference. If the jacks are spaced 24 in. the common difference as given on the square is 31.24 in Fig. 129 and 2 ft. 7¼ in. in Fig. 130.

To use the table proceed as follows:

1. Determine the center to center spacing of the jack rafters. This will be shown on the blue print or in the specifications. Locate this figure on the square.

2. Read the blue print and find the cut or pitch of the roof.

3. Find the number on the edge of the square blade that corresponds to the roof rise per foot of run.

4. Under this rise figure will be found the common difference in the length of jack rafters.

5. Convert decimal to fraction by using either the $\frac{1}{100}$ scale on the square (see Fig. 126), the edge of the square divided into tenths, or the decimal equivalent table on page 146.

10. Side (or Cheek) Cut of Jack Rafters.—(Figures 129 and 130.) The figures given on the square for the side cut of a jack rafter are based on the cut of the roof. Again, (as in previous square tables) each number on the linear measurement scale represents the rise in inches. *The run of* 12 *in. is always assumed.*

On some squares only one figure is given for the side cut; the other figure to use is 12. On some squares two numbers are indicated. The cutting is always done on the larger number. Two numbers given on a square may not be the same as two numbers shown for the same pitch, on another square. A comparison of the angles laid out by using the two sets of numbers will show them to be the same.

On the Stanley square only one figure is given; on the Nicholls square two figures are given. In both cases, however, the figures are selected under the number on the edge of the square that represents the roof rise.

To illustrate from the Stanley square (see Fig. 129):

What figures should be used on the square to give the correct side (cheek) cut for a jack rafter that is being framed for a 1/3 pitch roof?

Solution: A 1/3 pitch roof has an 8 and 12 cut. Under the figure 8 and in the space indicated by the heading at the end of the table will be found the figure 10. This number and 12 give the side cut; cut on 12. If the roof cut is 9 and 12 the side cut for the jacks will be 9 5/8 and 12; cut on 12.

Using the Nicholls square (see Fig. 130) what is the side cut for a jack rafter having a 3/8 pitch roof?

Solution: A 3/8 pitch has a 9 and 12 cut. Under the figure 9 will be found the figures 12 and 15; cut on 15. If the roof cut is 8 and 12 the side cut of the jack will be 10 and 12; cut on 12.

To use the table proceed as follows:

1. Determine the cut of the roof and select the rise figure on the edge of the square.

2. Under the rise figure read the number in the proper space.

Note: The Sargent square (see Fig. 132) also has a similar table for determining the side cuts of jack rafters. Two figures are given; the pitch in this case is determined by reading the heading at the end of the table.

11. Side (or Cheek) Cuts for Hip or Valley Rafters.—(Figures 129, 130 and 132.) The explanation as given for the side cut of a jack rafter is identical to the hip or valley rafter cheek

FIG. 132.

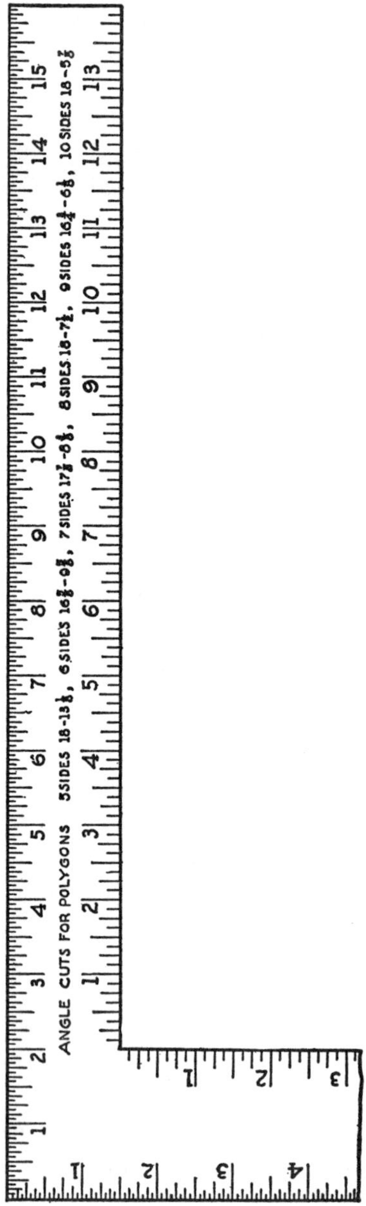

FIG. 133.

cut. A table will be found on the square that is read as described above for the jack rafters.

To use the table proceed as follows:

1. Follow all instructions as given for the jack rafter but use the hip or valley rafter table.

12. Sheathing Cuts.—(Figure 130.) 1. The sheathing cut figures on the Nicholls square are used to lay out the face or top cut required to make the board fit on the hip or valley rafter.

The set of numbers on the square is selected from the sheathing table on the basis of the cut of the roof. Each figure on the linear measurement scale (the outside edge of the body of the square) represents the rise of the roof.

To illustrate:

What is the sheathing cut for a 3/8 (9 and 12) pitch roof?

Solution: Under the figure 9 and in the lowest space will be found the figures 15 and 12. Cut on 12.

To use the table proceed as follows:

1. Determine roof cut from the blue print.
2. Select the rise in inches number on the edge of the square.
3. In the proper horizontal space and below the rise number read the sheathing cut figures. Be sure to cut on the *smaller* figure.

13. Polygon Miter Cuts.—(Figure 133.) The Sargent and Nicholls[1] squares each give a series of polygon miter cuts. Each polygon is in progressive order beginning with the pentagon (5 sides) and continuing up to 10 or 12 sides. The *smaller* number is used in cutting the miter.

To illustrate:

What is the miter cut for an octagon (eight sides)? From the table on the Sargent square (Fig. 133) the answer is 18 and 7½.

To use the table proceed as follows:

1. Determine the name of the polygon to be laid out, that is, hexagon (6 sides), heptagon (7 sides), etc.

[1] One of the Nicholls squares includes a complete table pertaining to all rafter cuts for an octagon roof.

2. Read the figures given for this polygon.
3. Cut on the smaller figure.

Note: Polygon cuts on the Nicholls square are all in terms of 24 in. as one of the numbers. The polygon cuts as given on page 131 are in terms of 12 in. as one of the numbers. The Sargent square figures use neither 24 or 12 in. *All* are correct, the angle being identical for any given set of numbers.

INDEX

J

L

M

O

P

R